Kings & Castles of the G.W.R.

O. S. NOCK

LONDON

IAN ALLAN

First published 1949
Second edition 1969

S.B.N. 7110 0071 9

Published by Ian Allan Ltd, Shepperton, Surrey and printed in the United Kingdom by
the Press at Coombelands Ltd, Addlestone, Surrey 571/EXX/569

Contents

Preface to the Second Edition

NEARLY 20 YEARS have passed since the first edition of this little book appeared and it is remarkable and a little sad to realise how completely the picture has changed in the meantime. I still live within sight and sound of the original main line of the G.W.R., and my monthly mileage on Western Region trains usually runs into four figures; but apart from occasional excursions by the privately-owned *Pendennis Castle* and *Clun Castle* the saga of the Great Western four-cylinder 4-6-0 working has ended.

It is, nevertheless, no story of outdated engines fading away that fills the years from 1949 to 1960. During that time Castles and Kings did some of the finest work of their entire lives, and three additional chapters have been necessary to complete the story. Much of it is written from first-hand experience: on the footplate, in the dynamometer car, and on the stationary testing plant. I am most grateful to the various officers of British Railways, Western Region who have accorded me so many privileges, and must mention particularly the late R. A. Smeddle, K. J. Cook, W. N. Pellow and H. E. A. White. Nor must I forget that great personality and splendid engineer S. O. Ell, who had charge of locomotive testing at Swindon. To travel with him was always an education.

O. S. NOCK

BATHEASTON
December, 1968.

Introduction

MOST OF US have our favourite locomotives. With some the choice may be due to the beauty of outline, with others personal association or admiration of their feats of haulage on the road. If long acquaintance was to be the deciding factor the Great Western four-cylinder 4-6-0s would be hot favourites among locomotives I have known, in fact, I might almost call them lifelong friends. As a very young schoolboy travelling daily from Mortimer to Reading I saw many of those wonderful engines, the Churchward Stars, new from the works; as an undergraduate, living in London, I saw the first Castles, and 'cut' college lectures to see the trains concerned in the 1925 exchange when the Castles were in competition with the Gresley Pacifics. In later years I grew to know these engines, Stars, Castles and Kings alike from that most intimate and revealing viewpoint—the footplate—and my admiration for them was enhanced.

I live on one of the principal main lines of the former G.W.R., and travel over it nearly every day of my life, and came to place almost complete reliance on the timekeeping of the long-distance trains I use. The Castles and Kings thus played a vital part in my daily comings and goings, and it is with all the greater pleasure that I looked out the gems of past performance which appear in these pages. It is a great story, and one that was by no means on the decline, for the Castles and Kings had been well maintained and could still repeat the feats of haulage that made them famous 40 years ago.

My thanks are due to Mr. Cecil J. Allen for permission to refer to runs timed by him in pre-war years. I am also very much indebted to Mr. A. V. Goodyear, a G.W.R. enthusiast of some 50 years standing, who not only made many of his own records available to me but also read the manuscript and made some valuable suggestions. Lastly, I am especially grateful to Mr. F. W. Hawksworth, Chief Mechanical Engineer of the Great Western Railway from 1941 to 1947, and later C.M.E. of the Western Region, through whose kindness I have many times been privileged to ride on the footplate of Castles, Kings, and indeed many other Great Western engines.

O. S. NOCK

BATH
March, 1949.

CHAPTER ONE

The Ancestry of the Castles

AT THE TURN of the century locomotive practice on the Great Western Railway was passing through a stage of great and rapid changes. It was only eight years since the Cornish expresses had run on the broad gauge, and were hauled out of Paddington by Gooch's famous eight-foot singles. In the years just before the final conversion from broad to standard gauge took place Great Western affairs were rather stagnant. The company could not press on with new developments while the tremendous engineering task of changing the gauge remained to be done; and even when the job was finished there was a good deal of leeway to be made up. But having thrown off the shackles of the broad gauge it would seem that the company became inspired with a new spirit: a spirit of boundless enterprise and enthusiasm, such as Brunel and Gooch infused into the building of the original line; and this spirit of enterprise was nowhere more apparent than in the Locomotive Department at Swindon. The traffic officers were certainly planning faster and more attractive services for the immediate future, but on looking back at those events it now seems that the Locomotive Department was planning even farther ahead.

The great G. J. Churchward was then chief assistant to the Locomotive, Carriage and Wagon Superintendent, William Dean, and in those early years a great deal of his attention was given to the design of boilers. The first engines embodying the ideas that he was gradually working out came in for a good deal of criticism on account of their rather unorthodox appearance. The Atbara class of 4-4-0s, built in 1900, had Belpaire fireboxes with the square top well above the level of the boiler, while the absence of a dome added to the rather gaunt appearance of the engines. Churchward was aiming at much higher steaming capacity, and increased power, by the use of higher pressures. This in turn involved higher temperatures, and to permit of these being used satisfactorily it was necessary to provide for the freest possible circulation of the water; otherwise one would get local 'hot spots,' with overheating of tubes and stays, and consequent leakage. The high raised Belpaire firebox of the Atbaras gave ample space above the water line and so greatly reduced foaming, while the use of a *tapered* boiler, as used on the famous City class 4-4-0s of 1903, gave a much increased water space between the fire tubes and the shell at the point where the boiler joined the firebox; here the heat is naturally greatest and circulation of the water most rapid. The tapered boiler, as well as being more efficient, had much more shapely appearance.

The boilers fitted to the City class, although providing steam for those engines to perform some wonderful feats of running, are of great importance to the student of locomotive practice, for they embody the great principles of design which have been followed so successfully in Great Western engines, the names of which are now household words among locomotive men. The first express passenger 4-6-0 built at Swindon, No. 100, preceded the Cities, and did not have a taper boiler in its earliest days; but in the second 4-6-0, No. 98, built in 1903, the forward ring of the boiler was parallel, and the rear one was tapered to the full width of the Belpaire firebox. In view of the great success of the Cities, there were many people at the time who questioned the need for such large engines as Churchward's new 4-6-0s; but this is where he was looking many years ahead. While it is true that when non-stop running between Paddington and Plymouth was first intro-duced in 1904 the load was limited to seven coaches, and could be managed comfortably by a 4-4-0, even over the very severe gradients of the South Devon line, the running of very much heavier trains was foreshadowed. Holiday traffic to resorts on the Torquay line was being fostered just as keenly as that to the Cornish Riviera, and with powerful engines available Torquay portions could be incorporated in the Cornish trains, and worked single-headed to Exeter, or Newton Abbot; afterwards the big engine was still available to tackle the South Devon line, with the reduced load.

The large engines had therefore not merely to possess the sheer strength necessary for ascending heavy gradients; they had also to be capable of developing maximum power at high speed. At that period in railway history speeds of 70 m.p.h. or more were usually attained while running down-hill under easy steam. Sixty m.p.h. on the level was the generally accepted hallmark of an express train, but Churchward aimed for something considerably higher. He set out to build engines capable of hauling heavy trains, for those days, of 300 to 350 tons at 70 m.p.h. on level track, and to sustain such speed for an hour or more. He had already produced a first-class boiler; it remained to design a 'front-end' that would use the steam economically. This he did by ensuring that the steam had the freest possible flow at all points in its circuit. He used piston valves of 10 inches diameter, with very large ports for both steam and exhaust; he arranged the valve gear with long travel, providing a 'slick' opening and closing of the valves and so avoiding loss of pressure due to restricted flow. Finally, by a clever adaptation of the Stephenson link motion he was able to give the engine increased power when starting, or when working with the cut-off well advanced in climbing a very heavy gradient. But even at this stage the resulting 4-6-0 did not look very much like a Castle or King.

Churchward, who in 1903 had succeeded Dean as Locomotive, Carriage and Wagon Superintendent, was very keenly aware of all that was happening in the locomotive world, not only in Great Britain, but on the Continent of Europe and in America; and one of the most important events of the day was the highly successful development of the De Glehn four-cylinder compound principle on certain of the French railways. The Nord, in particular, had some Atlantics that were doing wonderful work with the boat expresses between Paris and Calais, and before finally deciding on his future standard practice Churchward felt that a trial of the system ought to be made on the Great Western. The compound principle had certain quite definite theoretical advantages; so far, and particularly in British applications, those theoretical advantages seemed to be outweighted by the practical difficulties in building and running compound locomotives. But the De Glehn compounds seemed to be in a class by themselves, and so it was arranged for one of these engines to be purchased. This was the celebrated No. 102 *La France*, and in 1903 she was put to work on the G.W.R. in competition with Churchward's third 4-6-0 No. 171 *Albion*. In No. 171 the boiler pressure was raised to Swindon's highest yet, 225 lb. per sq. in., so as to be closely comparable to the 227 lb. per sq. in. of the French compound.

At this stage another important principle of Churchward's design should be emphasised. He used cylinders having a very long stroke in relation to the bore—18 in. diameter and 30 in. stroke. In this way he aimed at getting thermal efficiencies equal to those obtained with the two-stage expansion of compound engines. In any event the G.W.R. 4-6-0 No. 171 was put through very comprehensive trials against *La France*, and it soon became fairly clear that the Swindon-built engine was having very much the best of it. Churchward wished the trials to be conducted on the fairest possible basis, and at this stage it seemed that *Albion* might well derive some advantages over her rival by having extra adhesion weight due to six-coupled wheels. So *Albion* was temporarily altered to an Atlantic, and a further set of trials was run; but again she performed as well as *La France* if not better, while having the very important advantage of being much simpler to manage, and no doubt a good deal cheaper to construct and maintain.

But there was a point about *La France* that impressed Great Western men even more than her brilliant exposition of the compound principle; that was the very smooth riding and generally "sweet" action resulting from the division of the drive between the two coupled axles. Instead of applying all the power through one axle the loads on journals, big-ends, and so on were more evenly distributed; the reciprocating parts were better balanced, and her effect on the track was not so severe. It was felt that for continuous high-speed running a four-cylinder engine would show up to advantage, and to test the matter in the most practical way Churchward decided upon a further series of experiments. By 1905 the De Glehn system had been further developed in France, and two compound Atlantics of the latest design, similar to those used in the Orleans road, were purchased; these became G.W.R. No. 103 *President* and 104 *Alliance*. To compete with these engines Churchward built a 4-cylinder simple Atlantic, No. 40, which was given the name *North Star*. The drive was divided, as in the French engines; the boiler was the Swindon standard of the day, and in using cylinders 14$\frac{1}{4}$ in. diameter by 26 in. stroke the principle of using a stroke that was long in relation to the bore was carried even further than in *Albion*.

North Star proved a very successful engine, and she marks the true beginning of the four-cylinder era on the Great Western. But it is important to emphasise that four cylinders have only been used on locomotives designed for continuous running at high speed. For general service, and particularly in mixed traffic work, the two-cylinder engines with their simpler design, and their very effective layout of the Stephenson link motion, have proved more than equal to every demand. *North Star* was built as an Atlantic so that comparative tests might be carried out on as equal terms as possible against the larger French compounds; but once Churchward was satisfied that it was not necessary to adopt compounding in order to obtain high thermal efficiency the construction of four-cylinder 4-6-0 express passenger engines was commenced in 1907, with the justly-famous Star class. These engines had the Walschaert's valve gear, arranged inside, and in this respect differed from the Atlantic *North Star*; this latter engine had an unusual arrangement, and although similar in effect to Walschaert's was peculiar in having no eccentrics. The second component of the valve motion, which is normally obtained from an eccentric, or a return crank, was derived from a connection to the crosshead of the other inside cylinder. It was very compact, and gave excellent service for over 20 years, until, indeed *North Star* was rebuilt as a Castle.

In the Star class proper, engines 4001 to 4010, nearly all the ingredients of modern Great Western express passenger locomotive design were present. Experiments were already in progress with superheating, and after various trial apparatus had been made the 'Swindon No. 3' superheater was put on to the 21st four-cylinder 4-6-0 engine to be built, No. 4021 *King Edward*, completed in June, 1909. This proved to be the standard form of superheater, used subsequently in many hundreds of Great Western engines. The last appliance to be incorporated in the Churchward four-cylinder 4-6-0s was the top feed apparatus. As early as 1905 experiments were being made in this direction, but it was not until 1911 that finality was reached, and the apparatus applied generally to all types of standard locomotives built at Swindon. Its details, together with other interesting items of construction, are described in the next chapter. The introduction of superheating, in conjunction with the high steaming capacity of the standard taper boiler, enabled the power of the engines to be enhanced by using larger cylinders, and in 1913, beginning with engine No. 4041 *Prince of Wales*, the diameter of the cylinders was increased from 14¼ to 15 in. By the end of 1914 there were 61 engines of the Star class at work, *North Star* herself having been converted to a 4-6-0 in 1909 and renumbered 4000. The final batch of 12 engines, the Abbey series numbered 4061 to 4072, was built in 1922-3, and by that time the stage was almost set for the emergence from Swindon Works of the first of the engines we know so well today, the Castles.

CHAPTER TWO

The Castle Design and Some Early Test Results

BY THE AUTUMN of 1921 the Great Western Railway had recovered sufficiently from the effects of the First World War to restore passenger train services to full pre-war standards of speed. Once more the Cornish Riviera Express ran non-stop from Paddington to Plymouth, 225·7 miles in 247 minutes; four times a day the distance between Paddington and Bristol was covered non-stop in two hours, and there were many expresses running over the 110·5 miles between Paddington and Birmingham in two hours, even though this latter schedule usually included one intermediate stop. Most of this fine work was entrusted to Churchward's Star class four-cylinder

4-6-0s; at that time their general performance was unequalled by any other British locomotive class, and although Churchward himself retired at the end of 1921, his successor, Mr. C. B. Collett, built another series of Stars—the Abbey series of 1922-3—which differed from preceding ones only in certain details of construction.

Passenger traffic to the West of England was very much on the increase. The loading of the Cornish Riviera Express and other popular trains began to rise well above pre-war figures, and competent though the Stars were, the need began to be felt for a more powerful engine. There was another factor too which may, or may not, have influenced matters. In the spring of 1922 the Doncaster Works of the Great Northern Railway completed the first Pacific engine to the designs of Mr. H. N. Gresley, as he was then, and it was very soon apparent that these engines were not merely big in size and weight, but that they were capable of big things on the road. In the 'Railway Magazine' of February, 1923, Mr. Cecil J. Allen wrote: ' . . . now, at long last, a British type has appeared—granted, a much heavier, larger, and dimensionally more powerful engine—which will, I believe, give a Great Western 4-6-0 points and a beating. A bold claim indeed, but from my observations, one that is justifiable." But this challenge to Great Western supremacy did not remain unanswered for long, and in August, 1923, the first of a new series of four-cylinder 4-6-0s, No. 4073 Caerphilly Castle, was completed at Swindon Works.

The new design, although considerably enlarged, was based directly upon the Star; the layout of the chassis, and the spacing of the wheels was the same, but to provide greater power the diameter of the cylinders was increased from 15 inches to 16 inches. With the boiler pressure remaining at 225 lb. per sq. in., this gave a considerably increased tractive effort of 31,625 lb., at 85 per cent. boiler pressure, against 27,800 lb. in the Stars. It was also greater than the tractive effort of the much heavier Gresley Pacifics, which, also at 85 per cent. of boiler pressure, was 29,835 lb. To supply the larger cylinders of the Castles a very fine boiler was designed, entirely in the well-established Swindon tradition, but larger than that fitted to the Stars. The grate area was increased from 27·07 sq. ft. to 29·4 sq. ft., not quite in proportion to the increase in tractive effort, but it proved more than adequate for the heavier steaming required. Externally, the most distinctive features of the Castles was the roomy cab, with side windows, and canopy extending backwards over the tender footplate. The frames were lengthened to support this larger cab, so that a Castle was just one foot longer than a Star.

Another pleasing feature of these engines was that on them, for the first time since the war, the full Great Western livery was restored, with copper-capped chimney, polished brass safety valve cover, polished brass beading over the splashers, and the quietly dignified lining-out on the background of Brunswick green. From about midway through the First World War all G.W.R. engines had been painted plain green without any lining; the chimneys were painted black all over, and the safety valve covers were painted green. It will be appreciated with what enthusiasm the Castles were greeted in their array of polished brass and copper work, and as other express passenger engines went through the shops they too were restored to their pre-war splendour. As originally built the Castles had the standard 3500-gallon tender, with its low sides, and capacity for six tons of coal; while this was significant of the moderate coal and water consumption of Great Western express engines it looked a little out of proportion on such a massive 4-6-0 locomotive.

In detailed design, as in broad principles, Mr. Collett followed previous Swindon practice almost in its entirety. The standard bar-framed bogie was used; the general layout of the cylinders and valve gear was unaltered, and the boiler included the "Swindon No. 3" superheater and the usual top-feed device. The superheater is specially designed to afford easy maintenance; unlike the Schmidt and the Robinson it is possible to withdraw any group of elements without the need for moving any other group, whereas in the alternative types of superheater mentioned only the elements in the lowest row can be so withdrawn. The "top feed" device was developed as a result of much study and experiment. When the feed water is introduced into the boiler through the ordinary type of clack valve local cooling takes place, as the water in the boiler receives a cooler douche from outside. With top feed the water is fed into the boiler *through the steam*, and is therefore very much hotter when it comes in contact with the water already in the boiler; furthermore, with the Great Western arrangement the feed water after entering through the clack valves cascades down through a series of trays, and descends into the boiler in the form of fine spray.

The first batch of Castles, constructed in the autumn of 1923, consisted of ten locomotives, bearing names and numbers as follows:

4073	Caerphilly Castle
4074	Caldicot Castle
4075	Cardiff Castle
4076	Carmarthen Castle

4077	Chepstow Castle
4078	Pembroke Castle
4079	Pendennis Castle
4080	Powderham Castle
4081	Warwick Castle
4082	Windsor Castle

The Great Western used to advertise South Wales as the country of castles, and certainly in this series Welsh names predominate. It was not long after the construction of the first batch of Castle class engines that Their Majesties King George V and Queen Mary visited Swindon Works—April 28th, 1924. The Royal Train was worked by engine No. 4082 *Windsor Castle*, and in returning from the Works siding to Swindon station the engine was driven by the King, with the Queen and various high officers of the G.W.R. also on the footplate.

This pleasant occasion naturally served to increase popular interest in the Castle class engines, and in the same month engine No. 4074 *Caldicot Castle* was subjected to a very severe road trial between Swindon and Plymouth. The train was a special, with the dynamometer car attached, and the full load of 14 coaches, plus the dynamometer car, was run non-stop from Swindon to Taunton, via Bristol. Successive reductions of load took place, at Taunton and again at Newton Abbot, so as to make the test load correspond to the maximum tonnage taken in service conditions. This trial gave some very striking results. It showed that the Castle class engines were not only capable of sustaining a high output of power, but they sustained it most economically. *Caldicot Castle* worked this test train, weighing 480 tons behind the tender, at a steady speed of 65 m.p.h. on the long stretch of level road between Yatton and the approach to Taunton, and to do it she needed no more than 22 per cent. cut-off, with full regulator. Indicator diagrams were taken at frequent intervals by the testing staff riding in the specially constructed shelter round the front of the engine, and these showed that between 1300 and 1400 horsepower was being developed in the cylinders. The boiler was steaming very freely, and pressure was maintained at 225 lb. per sq. in. throughout this long spell of fast running.

Then, with a reduced load of 390 tons, a remarkable climb was made to Whiteball summit. Soon after starting the cut-off was set at 30 per cent., and the regulator opened to the full; in the easier stages of the ascent, between Norton Fitzwarren and Wellington, where the gradients vary between 1 in 369 and 1 in 174, speed was maintained at well over 50 m.p.h., and the indicated horsepower reached a maximum of 1600 when passing Wellington at 54 m.p.h. The engine was indeed going so well that no lengthening of the cut-off was needed to climb the really steep final pitch to Whiteball summit, where in three miles the gradient steepens from 1 in 90 to 1 in 86, and then to 1 in 80, before easing to 1 in 127 for the half mile through the tunnel and to 1 in 203 for the last 100 yds. to the summit. At the entrance to the tunnel speed had fallen to 36 m.p.h., but on the easier final pitch it rose to 38 at the summit signal box.

Equally fine work was done over other sections of the line, and when the test results came to be worked out the Castle class locomotives were clearly established as among the most economical engines which had run on British metals up to that time. First of all the capacity of the boiler proved unusually high; for every pound of coal fired roughly one gallon of water was evaporated, so although a great deal of very hard work was done the coal consumed was relatively low. A standard measure of coal consumption is the amount used in sustaining one horsepower at the drawbar of the tender for one hour; this measure is called a drawbar-horse-power-hour. On the fast run between Yatton and Taunton *Caldicot Castle* was maintaining 1000 horsepower on the drawbar at the back of the tender, while 1300 to 1400 horsepower was recorded by the indicators attached to the cylinders. The difference is the power absorbed in propelling the engine and tender.

Now at the time of these trials 4 lb. of coal per drawbar-horse-power-hour was a very common figure for good British locomotives; with some famous engines the consumption was 5 lb., and with others as much as 6 lb. But on that round trip from Swindon to Plymouth and back the consumption of *Caldicot Castle* worked out at no more than 2·83 lb. This, of course, was a wonderful result, and a great triumph for Mr. Collett and all those at Swindon who had taken a share in the designing and building of the Castle class engines. But it is most important to emphasise that this result was achieved with the best quality Welsh coal; practically all the coal that was shovelled through the fire door was actually burnt, very little being thrown away as sparks from the chimney. How big a difference coal can make was shown in some trials two years later on the L.M.S.R. on which the coal consumption of another engine of the Castle class worked out at 3·78 lb. per drawbar-horsepower-hour.

CHAPTER THREE

Fine Everyday Running

ALL TEN of the batch of Castles were originally stationed at Old Oak Common shed. One of them sometimes worked the relatively slow 10.45 a.m. from Paddington to Gloucester, and returned with what was even at that early date a famous train, the 2.30 p.m. up Cheltenham Flyer. It was already established as the Empire's Fastest Train, though the final run of 77·3 miles from Swindon to Paddington was then allowed as much as 75 minutes. With the load rarely reaching as much as 300 tons it was an easy task, not merely to a Castle, but almost equally to a Star or one of the two-cylinder Saint class 4-6-0s. Another turn from Old Oak worked at times by one or other of the first batch of Castles was the 9.10 a.m. two-hour express from Paddington to Birmingham; this involved a round trip of 246 miles, to Wolverhampton and back, and the return working was on the heavy "Northern Zulu", as the train leaving Birmingham at 3 p.m. was known among the G.W.R. staff. Although the time allowance from Snow Hill to Paddington was 2 hrs. 5 mins., this included two intermediate stops, and the final run of 67·5 miles from Banbury to Paddington in 70 minutes was one of the hardest duties on the Birmingham service.

But it was above all on the West of England main line that the reputation of the Castles was made. At the height of the holiday season the Cornish Riviera Express was run in two or more sections, and the only slip-coach portion was that for Weymouth, which was detached at Westbury. During the winter months, however, additional slip-coach portions were carried for Ilfracombe and Mine-head, detached at Taunton, and for the Torquay line, detached at Exeter. At the busiest times, that was just before the full summer service was introduced, and just after its conclusion, the formation of the Cornish Riviera Express reached a maximum of 14 coaches, and the gross load from Paddington was between 520 and 530 tons. This was reduced to 12 coaches, 450 tons, from Westbury; 10, or 385 tons, from Taunton, and usually to no more than 7 from Exeter, about 260 tons behind the tender. The load was thus reduced in some conformity with the increasing severity of the gradients. The schedule throughout was very fast, requiring an average speed of 58·8 m.p.h. for the 95·6 miles from Paddington to Westbury; 56·2 m.p.h. for the 47·3 miles from Westbury to Taunton;

Process of Evolution

Atbara: with domeless boiler and high raised Belpaire firebox [*LPC*

The French influence: de Glehn 4cyl compound Atlantic No 104 *Alliance,* as later fitted with Swindon standard boiler [*M. W. Earley*

Caerphilly Castle, as originally built in 1923, with small tender [*British Rail*

The three generations: King, Castle and Star [*W. J. Reynolds*

Early Days of the Castles

In June 1925, No 4083 *Abbotsbury Castle,* was photographed on the Royal Train bound for Newbury Racecourse

[*M. W. Earley*

No 4082 *Windsor Castle* in state, working the GWR Royal Train at the Railway Centenary celebrations at Darlington 1925

[*F. R. Hebron*

Over the crossover at the old Twyford East Signal Box, No 4094, still with small tender, comes with the up Plymouth Ocean Mails Special   [*F. R. Hebron*

Windsor Castle again, minus her safety valve cover, working the 1 30pm Paddington to Penzance express near Hayes   [*F. R. Hebron*

The protagonists in the 1925 locomotive exchange: GWR No 4079 *Pendennis Castle* and LNER No 4475 *Flying Fox* alongside at Kings Cross Top shed　　　　[*W. J. Reynolds*

The locomotive rebuilt from the pacific *The Great Bear* was *Viscount Churchill* as Castle class, but still numbered III Here it is at the head of the "Cheltenham Flyer" [*Fox Photos*

59·6 m.p.h. for the 30·8 miles from Taunton to Exeter—a most difficult booking with a heavy train requiring to be worked up the stiff grade to Whiteball tunnel. Then with the load reduced to the minimum there was an easy allowance of 24 minutes for the 20·2 miles to Newton Abbot, taking into consideration the moderate speed necessary on the curving stretches along the coast from Starcross to Teignmouth, and finally an average of 43·3 m.p.h. for the last 31·8 miles from Newton Abbot to Plymouth, a timing that did not leave a great deal of margin due to the exceptionally severe character of the road.

But, above all, the run had to be made non-stop. There was no respite, and if an engine should go a little shy for steam, fire-cleaning or other remedial operations had to be carried out at speed, unless serious loss of time was to be incurred. In recent years, when the quality of coal has been so variable, I have been on the footplate many times when the locomotives have been in trouble for steam, and I know only too well how very welcome a brief stop can be, even if it means hard running afterwards. The regularity with which the Cornish Riviera Express was run by the Castles was a fine tribute to the design, and to the work of the drivers and firemen. An excellent run by engine No. 4079 *Pendennis Castle* was recorded from the footplate by Mr. Cecil J. Allen in 1924, and from details published in the *Railway Magazine* some interesting points may be mentioned. The timekeeping was accurate, with a minute or so in hand all the way; the actual averages over the sections previously quoted were 60 m.p.h. from Paddington to Westbury, with 530 tons; 58 m.p.h. from Westbury to Taunton, with 450 tons; and 57·8 m.p.h. from Taunton to Exeter despite a slack through Whiteball tunnel. Exeter was passed 2½ minutes early, and with 275 tons the difficult final timing from Newton Abbot to Plymouth was slightly improved upon, even though two slowings for permanent-way work were in progress. Plymouth was reached three minutes early.

Now first of all in making the fast run from Paddington to Exeter, 173·7 miles in 176½ minutes, the main valve of the regulator was open for less than two-thirds of the time, for 106 minutes, to be exact, out of 176. For the rest of the time only the small auxiliary port, or "first valve" as the enginemen call it, was used. It was with the "first valve" that all the fast downhill running was performed, such as an average of 73·2 m.p.h. over the 19 miles from Pewsey to Milepost 94¼ (just before Westbury) or the average of 71·5 m.p.h. for the 13·6 miles from Tiverton Junction to Cowley Bridge Junction in the approach to Exeter. The longest spell of

CORNISH RIVIERA EXPRESS—May 2nd, 1925

Engine—4074 *Caldicot Castle*. Driver Rowe, Fireman Cook (Old Oak Common), Load to Westbury: 498 tons tare, 530 tons full; Load to Taunton: 426 tons tare, 455 tons full; Load to Exeter: 363 tons tare, 390 tons full; Load to Plymouth: 292 tons tare, 310 tons full.

Dist. miles					Schd. min.	Actual min. sec.		Speeds m.p.h.
0·0	PADDINGTON	..	..	..	0	0	00	
1·3	Westbourne Park	..	..	..		3	15	
5·7	Ealing	..	..	..		9	08	
9·1	Southall	..	..	..	11	12	42	59
13·2	West Drayton	..	..	..		16	41	65½
16·2	Langley	..	..	..		19	26	66½
18·5	SLOUGH	..	..	..	20	21	26	68
24·2	Maidenhead	..	..	..	25½	26	36	66½
31·0	Twyford	..	..	..	31½	32	49	66½
36·0	READING	..	..	..	37	37	25	
37·8	Southcote Jc.	..	..	..		39	31	45
44·8	Aldermaston	..	..	..		46	20	62
46·7	Midgham	..	..	..		48	15	61
49·6	Thatcham	..	..	..		51	06	60½
53·1	NEWBURY	..	..	..	56	54	38	59½
58·5	Kintbury	..	..	..		60	05	60½
61·5	Hungerford	..	..	..		63	12	55½
66·4	Bedwyn	..	..	..	69½	68	17	57
70·1	Savernake	..	..	..		72	25	46
75·3	Pewsey	..	..	..		77	25	66
81·1	Patney	..	..	..		82	16	70½
86·9	Lavington	..	..	..		87	00	77½
91·4	Edington	..	..	..		90	38	69
95·6	WESTBURY	..	..	..	97½	94	40	30
101·3	FROME	..	..	..		101	40	30
106·6	Witham	..	..	..		108	03	51½
108·5	Brewham Summit		..	..		110	24	46
115·3	Castle Cary	..	..	..	120	116	18	75 (max.)
								60
120·2	Keinton Mandeville		..	..		120	26	72½
125·7	Somerton	..	..	..		125	28	64
131·0	Curry Rivel Jc.	..	..	..		130	09	72
137·9	Cogload Jc.	..	..	..	143	136	06	55
								69
142·9	TAUNTON	..	..	..	148	140	30	64
144·9	Norton Fitzwarren	..	..	..		142	22	67½
150·0	Wellington	..	..	..		147	17	58
152·7	Milepost 173	..	..	..		150	42	41
153·8	Whiteball Box	..	..	..		152	12	45
158·8	Tiverton Jc.	..	..	..		156	46	75
161·1	Cullompton	..	..	..		158	38	76½
166·5	Silverton	..	..	..		162	56	72½
170·2	Stoke Canon	..	..	..		166	00	74
173·7	EXETER	..	..	..	179	169	10	30 (slack)
178·4	Exminster	..	..	..		174	40	70½
182·2	Starcross	..	..	..		178	00	60
188·7	Teignmouth	..	..	..		184	54	
193·9	NEWTON ABBOT		..	..	203	190	25	
						p.w.	slack	30
195·0	Aller Jc.	..	..	..		192	25	

197·7	Dainton Box..	..	..	..	209½	197	40	23½
202·5	TOTNES ..	..	..	..	215½	203	00	60 (max.)
205·3	Tigley Box ..	..	..	..		206	55	29
207·1	Rattery Box	..	..	..	223	210	25	35
209·4	Brent ..	..	..	..	225	213	28	60
211·6	Wrangaton ..	..	..	..		215	40	52½
219·0	Hemerdon Box	..	..	..	237	223	10	64
221·7	Plympton ..	..	..	..		225	41	70½
						p.w. slack		20
223·8	Laira Jc. ..	..	..	..		228	30	
225·7	PLYMOUTH	..	..	..	247	231	58	

continuous work on the main regulator valve was in the first 70 miles, from Paddington out to Savernake. Here time was kept almost to the second, and this stretch covered in 73 min. 25 sec. The engine was being worked reasonably hard here, with 26 to 30 per cent. cut-off; but the quality of performance can be judged from such feats as a sustained maximum speed of 69 m.p.h. on the level, and an uphill average of 58·3 m.p.h. throughout the long rise of 32·3 miles from Southcote Junction up the Kennet Valley, to Savernake. West of Exeter the main valve of the regulator was used for only 16 minutes out of the total of 67½ minutes running time from Exeter to Plymouth. Then, it is true, the going was really hard in two short spells up the Dainton and Rattery inclines, with cut-off advanced to a maximum of 42 per cent. on the first, and 41 per cent. on the second. The fine work was performed, as usual with the Castles, most economically, and the water consumption was no more than 30 gallons per mile.

This was a normal timekeeping performance. In view of the relative ease with which a great deal of the journey was performed it might well be asked what a Castle could do if driven with the idea of making a *really* fast time. This question is to some extent answered by the performance of engine No. 4074 *Caldicot Castle* in the exchange trials of 1925, when she reached Plymouth in 231 min. 58 sec.—15 minutes early—on May 2nd, 1925. As this run is of historic interest I have prepared the log opposite. It will be seen that there was very little improvement over the work of *Pendennis Castle* as far as Westbury, thus emphasising the difficult nature of this opening schedule with the full load; but after that the average speeds were 61·9 m.p.h. instead of 56·2 booked, from Westbury to Taunton; 64·5 m.p.h. instead of 59·6 from Taunton to Exeter, and 46 m.p.h. instead of 43·3 booked from Newton Abbot to Plymouth. The latter timing was especially good, as the load from Exeter was 8 coaches instead of the usual 7—310 tons behind the tender.

But even without the stimulus of test conditions the drivers and firemen of the Castles used to make equally notable times. A later engine of the class, No. 5009 *Shrewsbury Castle*, made some exceedingly fine running to Exeter on a West of England express that conveyed no slip portions and had a load of 14 coaches, 452 tons tare and at least 490 tons gross. The haulage effort required was therefore only a little easier than that of the 14-coach Cornish Riviera Express as far as Westbury, considerably harder than the usual 450-ton load onwards to Taunton, and a good 100 tons more over the stiffest piece of all, from Taunton to Exeter. In spite of this, and still more in spite of some heavy slacks for adverse signals, the 173·7 miles from Paddington to Exeter were covered in 181½ minutes, start to stop. When allowance is made for the signal checks the net time works out at 172½ minutes—a wonderful achievement.

The mention of engine No. 5009 *Shrewsbury Castle* has carried the story forward to the second and third batches of Castles. The second batch consisted of 10 locomotives 4083-4092, and was completed at Swindon in May, 1925; the third batch consisted of 20 locomotives 4093-4099 and 5000-5012, and this was completed in July, 1927. In these later engines no changes were made from the original design, in fact construction of the class went on until 1939 with certain very minor changes in engines numbered from 5013 onwards. In addition to the new engines built in 1925-7 the stud of Castle class engines was strengthened by the re-building of some earlier engines of the Star class, and by the virtual scrapping of the famous Pacific engine *The Great Bear*, and using her frames, wheel centres and motion as the basis for a new "No. 111." I was rather sorry that the fine old name was not preserved on this re-built engine; instead she was named *Viscount Churchill*, after the Chairman of the Company.

The new engines in these two batches were:

4083	Abbotsbury Castle	4095	Harlech Castle
4084	Aberystwyth Castle	4096	Highclere Castle
4085	Berkeley Castle	4097	Kenilworth Castle
4086	Builth Castle	4098	Kidwelly Castle
4087	Cardigan Castle	4099	Kilgerran Castle
4088	Dartmouth Castle	5000	Launceston Castle
4089	Donnington Castle	5001	Llandovery Castle
4090	Dorchester Castle	5002	Ludlow Castle
4091	Dudley Castle	5003	Lulworth Castle
4092	Dunraven Castle	5004	Llanstephan Castle
		5005	Manorbier Castle
4093	Dunster Castle	5006	Tregenna Castle
4094	Dynevor Castle	5007	Rougemont Castle

5008	Raglan Castle	5011	Tintagel Castle
5009	Shrewsbury Castle	5012	Berry Pomeroy Castle
5010	Restormel Castle		

To complete this reference to their early work there is a very fine piece of running I recorded with engine No. 4088 *Dartmouth Castle* on the 6.10 p.m. express from Paddington to Wolverhampton. This was always a fairly heavy train of about 430 to 440 tons out of London, but slip portions were detached at Bicester and at Banbury, which reduced the load considerably by the time the difficult 26-minute run from Leamington to Birmingham had to be tackled. But on this occasion, just before the Whitsun holiday, the train was divided, and the first section included no slip portions; instead, an extremely heavy load of 475 tons had to be conveyed right through to Wolverhampton. This, of course, was only one coach more than the regular train as far as Bicester; but while any good driver would be prepared to cope with one extra coach, the prospect of two more than usual from Bicester, and *four* more from Banbury was a different matter altogether. From the very start, however, the running was grand. We sustained 51 m.p.h. up the last stages of the long and rather trying climb from Denham to Beaconsfield, on a gradient of 1 in 254; High Wycombe, Ashendon Junction, and Bicester were all passed dead on time, and then a really big effort was made over the concluding stages to Leamington.

The long rise of $5\frac{1}{2}$ miles at 1 in 200 to Ardley tunnel was cleared without speed falling below $52\frac{1}{2}$ m.p.h., and from the water troughs at Aynho Junction the rise of 11·2 miles to Claydon Crossing, beyond Cropredy, was taken at an average speed of 64 m.p.h., after which a fast though not exceptional descent into Leamington followed. The maximum speed here was 78 m.p.h., whereas on some occasions peak rates of nearly 90 m.p.h. were touched. But in any event, this remarkable work—and particularly that performed uphill—was such as to bring us into Leamington $1\frac{1}{4}$ minutes early, 89 min. 50 sec. for the 87·3 miles from Paddington. But with this great load we should have lost at least 3 minutes onwards to Birmingham. The timing of 26 minutes for this run of 23·3 miles was designed for a load of about 300 tons, not 475; in the 4 miles of Hatton bank, averaging 1 in 110, we fell from 45 to $32\frac{1}{4}$ m.p.h., and although we did very well onwards to Birmingham the load was too great for the section timing to be observed, and with the additional handicap of a slight check for permanent-way work near Tyseley, the time to Birmingham was $29\frac{1}{2}$ minutes. But of this

6.10 p.m. PADDINGTON—BIRMINGHAM

Engine—4088 *Dartmouth Castle*
Load—440 tons tare, 475 tons full
Driver—A. Taylor (Stafford Road)

Dist. Miles					Schd. min.	Actual min. sec.		Speeds m.p.h.
0·0	PADDINGTON	..	..	..	0	0	00	
						signals		
3·3	Old Oak Common West			..	7	7	20	
7·8	Greenford	..	..	..		12	55	57½
10·3	Northolt Jc. ..	..	..	..	15½	15	35	54¼
14·8	Denham	..	..	..		20	05	63½
17·4	Gerrards Cross	..	..	..		22	55	52½
21·7	Beaconsfield	..	..	..		27	50	51
24·2	Tylers Green	..	..	..		30	05	70½
26·5	HIGH WYCOMBE	..	..	..	32	32	15	50 (slack)
28·8	West Wycombe	..	..	..		35	05	47
31·5	Saunderton	..	..	..		38	45	42
34·7	PRINCES RISBOROUGH		..	..	41	42	30	70½
40·1	Haddenham	..	..	..		46	40	83½
44·1	Ashendon Jc.	..	..	..	49	49	40	58 (slack)
47·4	Brill ..	..	..	..	..	52	55	
50·4	Blackthorn ..	..	..	..		53	35	70½
53·4	BICESTER	..	..	..	58	58	15	62
57·2	Ardley	..	..	..	..		62	15	52½
								69
62·4	Aynho Jc. ..	..	..	..	67	67	15	64¼
67·5	BANBURY ..	..	..	..	72	71	55	67
71·1	Cropredy	..	..	..	..		75	15	57½
76·2	Fenny Compton	..	..	..		80	15	76½
81·2	Southam Rd.	..	..	..		84	15	74
83·6	Fosse Rd. Box	..	..	..		86	10	78
87·3	LEAMINGTON	..	..	..	91	89	50	
0·0					0	0	00	
2·0	WARWICK ..	..	..	..		3	45	45
6·2	Hatton	..	..	..	..		10	15	32
10·4	Lapworth	..	..	..	..		15	20	58
12·9	Knowle	..	..	..	..		18	05	54½
16·3	Solihull	..	..	..	..		21	40	60½
20·1	Tyseley	..	..	..	..		25	00	68
						p.w. slack		
22·2	Bordesley ..	..	..	..		27	10	
23·3	BIRMINGHAM	..	..	..	26	29	30	

loss the driver regained 1¾ minutes by making a smart run through the Black Country and covering the 12·6 miles from Birmingham to Wolverhampton in 18¼ minutes. Thus the 3 minutes net loss between Leamington and Birmingham was exactly balanced, as 1¼ minutes had been gained by the very fine run from Paddington to Leamington.

CHAPTER FOUR

The Building of the Kings

DESPITE the excellent performance of the Castles, and their outstanding success in the exchange trials of 1925 with the L.N.E.R., and of 1926 with the L.M.S.R., the demands of Great Western express passenger train service were soon to require something still better. The Company was not content with restoring the standards of speed prevailing before the outbreak of the First World War, in 1914; further accelerations were planned, while the growth of the travel habit, and the increased number of people travelling relatively long distances for annual holiday, made it fairly clear that the accelerated timings, when they were put into effect, would have to be maintained with very much heavier trains. In the meantime the performance of the Castles on the foreign lines in turn profoundly influenced locomotive practice on the L.N.E.R. and on the L.M.S.R. The story of the exchanges has been told very fully by Mr. Cecil J. Allen,* and I need not say more at this stage than that in 1925 No. 4074 *Caldicot Castle* on the G.W.R., and 4079 *Pendennis Castle* on the L.N.E.R. between them claimed most of the honours in the trials against the Gresley Pacifics, and that even on *Caldicot Castle's* record trips, when she brought in the "Limited" 15 minutes early, her coal consumption was less than that of the Pacific making the same journey in a quarter of an hour longer time.

The L.M.S.R. then had no engine that could compete on level terms with No. 5000 *Launceston Castle*, and the successful running of this latter engine led, by a rather circuitous chain of events, to the building of the Royal Scots. After these trials both the Northern lines became converted to the use of high boiler pressures, and long travel valves, and the efficiency and speed-worthiness of their engines improved accordingly. The fourth British company, the Southern, had adopted long travel valves at a much earlier date; in fact one of its constituents, the South Eastern and Chatham, was building engines very much in the Swindon tradition as early as 1917. That was not altogether surprising, as Mr. Maunsell's chief assistant, G. H. Pearson, was an ex-G.W.R. man, and there were several other Swindon men on the staff at Ashford.

So the Great Western had influenced each of the other companies in turn; but at the time when the Castles were extending their

* In "The Locomotive Exchanges 1870-1948": Ian Allan Ltd.

sphere of influence to the Scottish border the Southern Railway completed at Eastleigh Works a new express passenger engine, which by a considerable margin wrested from the Castles the claim to be the most powerful passenger locomotives in Great Britain. In the autumn of 1926 the *Lord Nelson* was finished, and the nominal tractive effort of this fine engine was 33,500 lb., at 85 per cent. boiler pressure, against the 31,625 lb. of the Castles. But the honour soon passed back to the Great Western, and in no half measure, either, for in the summer of 1927 engine No. 6000 *King George V* was completed at Swindon Works—a new four-cylinder 4-6-0 having a nominal tractive effort of no less than 40,300 lb.

In this remarkable design all the well-established features of Swindon practice were carried a notable stage forward. The boiler pressure was increased for the first time since Churchward's day, from 225 to 250 lb. per sq. in., the first time so high a pressure had been used on a British locomotive. Loading gauge restrictions precluded any appreciable increase in the cylinder diameter, but enlargement took place precisely in the manner that Churchward taught, by making the stroke long in relation to the bore. Since the construction of the first Stars with $14\frac{1}{4}$ in. diameter cylinders having 26 in. stroke, and therefore a ratio of 1 to 1·82, this ratio had decreased to 1 to 1·73 in the later Stars with 15 in. cylinders, and to 1 to 1·62 in the Castles. In the Kings, which have cylinders $16\frac{1}{4}$ in. diameter by 28 in. stroke, the ratio was increased again to 1 to 1·73. The adhesion weight was increased to no less than $67\frac{1}{2}$ tons, but the only point of doubt, to the outsider, was whether the boiler would be able to supply sufficient steam to make the very high nominal tractive effort a practical reality. For although the boiler and firebox was considerably larger than that of the Castles, it was still a good deal smaller than that of the Gresley Pacifics.

In the latest development of this latter class, No. 4480 *Enterprise*,* rebuilt with a boiler carrying a pressure of 220 lb. per sq. in., as compared with 180 lb. in the engines that competed with *Caldicot Castle* and *Pendennis Castle* in 1925, the total heating surface was 3442·6 sq. ft. against 2514 sq. ft. in the *King George V*; the respective grate areas were 41·25 sq. ft. and 34·3 sq. ft., and the superheater heating surface was 706 sq. ft. on the Pacific, against 313 on the King. The whole design of the King class was most compactly arranged, and it was remarkable that such a very powerful engine could be placed upon a 4-6-0 chassis, and weigh no more than

* Later No. 60111

89 tons. The Gresley Super-Pacific *Enterprise* weighed 96¼ tons without her tender. It remained to be seen if the new G.W.R. engines were as effective in service as their high nominal tractive effort implied.

It was the famous Locomotive Engineer of the Great Northern Railway, H. A. Ivatt, who said that the power of a locomotive depended upon its capacity to boil water; this *dictum* can be taken one stage farther back, and be related to the capacity to *burn* coal. I have italicised the word "burn" because it is this capacity which lies at the heart of Great Western locomotive performance. The Swindon fireboxes, whether on Castles, Kings or on the various two-cylinder engines, had a wonderful capacity for *burning coal*. With the good Welsh grades the fire was built up thick, packed in the rear corners, and piled in just above the bottom level of the firedoor. It swells into a single incandescent mass with a crust of partly black coal on the top, and the steam-raising capacity of such a fire seems unlimited.

But one must not give all the credit to the coal! Successive generations of enginemen, brought up in the Great Western tradition, have become most expert in getting the very best out of these engines. One could not help noticing how rare it was, even with coal other than Welsh, for Great Western engines to smoke heavily. In the exchange trials of 1948 I travelled from Leeds to King's Cross behind engine No. 6018 *King Henry VI*; at times there was a slightly grey tinge to the exhaust, but never any real black smoke. And this brings us back to some of the earliest exploits of the Kings. On her first trip with the Cornish Riviera Express, the pioneer engine No. 6000 *King George V* excited comment in that she was fitted with the Westinghouse air brake; the engine was shortly to be shipped to America to take part in the centenary celebrations of the Baltimore and Ohio Railroad. This exhibition was open from September 24th to October 15th, 1927, and was visited by more than a million and a quarter people.

A New York correspondent of the *Railway Magazine* who visited the exhibition, wrote: "It was gratifying to note that the Great Western Railway of England had easily the most popular exhibit in that triumph of British locomotive design—the *King George V*. It was especially notable that throughout the pageant, and, indeed throughout the afternoon and evening, the *King George V* never showed so much as one wisp of black smoke, much in contrast to all the Canadian and American engines, which smoked freely pretty well all the time. This was the more remarkable in that the Great Western engine was using American coal, and seems to point to the fact that the latter has a firebox more efficient than the American and Canadian wide type. I found Driver Young— who in view of his exploits on the L.N.E.R. and the Western Division of the L.M.S. must surely be the most experienced express driver in Britain—a quiet-spoken, unassuming man, but with a quaint dry humour. I may add that it took me a solid half-hour to get on the footplate of *King George V* when the pageant was over!" I should add that he was Driver W. Young, of Old Oak Common shed, who did so well with *Pendennis Castle* on the King's Cross-Doncaster trials of 1925 and with *Launceston Castle*, between Euston and Carlisle in 1926.

At home public interest in Great Western locomotives was still further enhanced, when on November 3rd, 1927, the first of a series of special 5/- excursions was run from Paddington to Swindon, inclusive of a conducted tour round the Works. So many prospective passengers applied that the sale of tickets had to be suspended two days before the trip, even though the train was run in duplicate. These trips proved so popular that they became a regular feature of the Great Western excursion programme, one of the advertised attractions being that the trains would be hauled between Paddington and Swindon by engines of the King class. The first day set the standard for the whole series of excursions, both trains were worked from Paddington to Swindon, 77·3 miles, in less than even time, and on the return trip one of the specials made a very fast run, covering the distance to Paddington in 68½ minutes, start to stop.

The first batch of Kings consisted of 20 engines, named after the Kings of England, as follows:

6000	King George V	6010	King Charles I
6001	King Edward VII	6011	King James I
6002	King William IV	6012	King Edward VI
6003	King George IV	6013	King Henry VIII
6004	King George III	6014	King Henry VII
6005	King George II	6015	King Richard III
6006	King George I	6016	King Edward V
6007	King William III	6017	King Edward IV
6008	King James II	6018	King Henry VI
6009	King Charles II	6019	King Henry V

There are several additional points about the design of these famous engines before we pass on to a serious consideration of their performance on the road. The bogie, which is spring controlled, was of an entirely new design, having outside bearings on the leading axle and inside bearings on the rear axle. It was desired to have independent springing for each of the bogie wheels, and to do this it was necessary to place the springs on the leading wheels outside the bogie frame, so as to clear the inside cylinders. The springs on the rear bogie wheels are, conversely, placed inside, so as to clear the outside cylinders.

The Kings were the first Great Western engines to have the fine-looking 4000-gallon tenders. I have previously mentioned that the old standard 3500-gallon tender looked a little out of place on the Castles, but the new type was subsequently fitted to all engines of that class, and also to many of the Stars.

CHAPTER FIVE

Some Great Runs with the Kings

ONE OF THE happiest features to be recorded of the Kings is the rapidity with which they found their true form in the most arduous express passenger service. On July 20th, 1927, No. 6000 *King George V* worked the Cornish Riviera Express for the first time; the day was a Wednesday, and the train carried its normal mid-week load for the summer service—12 coaches to Westbury, and 10 beyond, with no slip portions detached at Taunton and Exeter. As far as Newton Abbot the task was a fairly easy one compared to the full winter load hitherto worked by the Castles; but the 10-coach train of 338 tons tare would have required pilot assistance beyond Newton Abbot. Here *King George V* showed complete mastery on such exceptional inclines as Dainton and Rattery, where the maximum tonnage permitted to a Castle, or with one of the modern Counties, was 315 tons. The load for the Kings west of Newton Abbot was fixed at 360 tons, and by their introduction a good deal of double-heading was avoided. There is little doubt, too, that the working of the "Limited" between Paddington and Westbury with the 14-coach winter load of about 490 tons tare imposed something of a task upon the Castles; in favourable running conditions they could manage it excellently, but the continuous use of cut-off between 25 and 30 per cent. all the way from Paddington to Savernake did not leave much margin to cope with bad weather, or for recovery from engineering slacks.

Thus the additional tractive power of the Kings was in particular demand in the early stages of the West of England run, and equally so west of Newton Abbot. It was very soon evident that the Kings had a comfortable mastery over the job on these two stretches.

Mr. Cecil J. Allen timed two runs from the footplate, on one of which engine No. 6011 *King James I*, with a load of 492 tons tare, 525 tons full, sustained 70 m.p.h. on the level while working on no more than 18 per cent. cut-off, and moreover made an average of 60½ m.p.h. for the 28 miles up the Kennet valley on cut-offs varying between 18 and 20 per cent. With the regulator full open this represented normal fast running, without making any unduly heavy demands upon the boiler. For long periods the pressure remained steady at 245 lb. per sq. in., and only for two brief periods on the climb to Savernake was any reading as low as 235 lb. per sq. in. recorded.

The confidence of the operating authorities in the capacity of the Kings was shown by the acceleration of the Cornish Riviera Express in the autumn of 1927. There was no waiting to see how the Kings managed the full 14-coach train; the schedule was cut to the even four hours from the outset. As can well be imagined, only a very little was taken out of the Paddington-Westbury time; the old and new timings were as follows:

Miles						Old Times	New Times
0·0	Paddington	..	..	..	..	0	0
36·0	Reading	..	..	..	..	37	37
66·4	Bedwyn	..	..	..	..	69½	68½
95·6	Westbury	..	..	..	..	97½	96
115·3	Castle Cary	..	..	..	..	120	118
142·9	Taunton	..	..	..	..	148	144½
173·7	Exeter ..	..	..	..	..	179	174½
193·9	Newton Abbot		..	..	..	203	198½
202·5	Totnes..	..	..	..	..	215½	210½
209·4	Brent ..	..	..	..	..	225	219½
225·7	Plymouth	..	..	..	..	247	240

The accelerated schedule did not unduly tax the Kings and on Mr. Allen's footplate journey with No. 6011, Exeter was passed in 176 minutes, after three very severe slacks for relaying; the net time was 167¾ minutes—nearly seven minutes inside schedule.

During the summer service, when only one slip portion was carried, the schedule was adjusted so as to give more time between Newton Abbot and Plymouth. This also made allowance for those occasions when the load was more than 360 tons tare, and a stop had to be made for pilot assistance. Then the booked passing time at Newton Abbot was 195½ minutes, but it is remarkable that of the three minutes cut from the timing to this point from Paddington, two were taken from the already severe 96 minutes allowance out to Westbury. The extent to which the Cornish Riviera Express loaded during the summer can be judged from two runs published in the *Railway Magazine* for March, 1933, when 15-coach trains were taken out of Paddington, reduced to 12 coaches in each case after Westbury. The allowance of 94 minutes for the 95·6 miles from Paddington to Westbury was kept with a comfortable amount in hand on both of these remarkable runs; engine 6028, then *King Henry II*, with 555 tons passed through in the wonderful time of 91 min. 30 sec., while *King Henry IV*, with 560 tons, took 92 min. 55 sec.

The first of these two runs was, throughout to Newton, a magnificent performance, and an abbreviated log is given herewith. Speed rose to 70½ m.p.h. on the level at Slough; the average from Southcote Junction to Bedwyn up the gradual rise of Kennet Valley was 61 m.p.h., and the minimum speed over the summit, after the final three miles at 1 in 175, 1 in 183, 1 in 145 and 1 in 106, was exactly 50 m.p.h. Then came a fast descent to Westbury, with a maximum speed of 82 m.p.h. at Lavington. At the time the track layout at Taunton station, and in the northern approaches, was being completely remodelled, and with some severe checks in store, the driver of No. 6028 now began to get in hand as much time as possible, to offset their effect. The reduction of load to 440 tons at Westbury was a considerable help, and the engine was worked to such good purpose that Athelney, 134·9 miles from Paddington, was passed in 128 min. 10 sec., enough to pass Taunton six minutes early, with normal speed on the approach lines.

But the checks were so severe that all this time was swallowed up, and a minute more as well. Indeed, the 10 miles from Athelney to Norton Fitzwarren took 16 min. 40 sec. But very fine work followed, with speed rising to 52½ m.p.h. up the rise to Wellington, and not falling below 39½ m.p.h. at Whiteball summit. This was considerably faster than *Caldicot Castle* performed with her 390-ton train in the dynamometer car test run of 1924. With high speed down to Exeter, with a maximum of 79 m.p.h., the total time for this 173·7 miles from Paddington was only 173 min. 55 sec., and the net time 167 minutes. Thus, despite this very heavy load, the net average speed from London was 62·4 m.p.h.; furthermore, by very brisk running round the coastal section, the loss of time occasioned by the Taunton delays was entirely wiped out, and Newton Abbot was reached in 193 min. 25 sec. for the 193·9 miles from Paddington.

Here the train actually stopped in two minutes less than the

CORNISH RIVIERA EXPRESS

Engine—6028 *King Henry II**

Load to Westbury   514 tons tare   555 tons full
        to Newton Abbot 409 tons tare   440 tons full

Dist. miles					Schedule min.	Actual min. sec.		Speeds m.p.h.
0·0	**PADDINGTON**	..	..	..	0	0	00	
1·3	Westbourne Park	..	..	..		3	00	
5·7	Ealing	..	..	..		9	05	
9·1	Southall	..	..	..	11	12	40	59
13·2	West Drayton	..	..	..		16	40	
18·5	**SLOUGH**	..	..	..	20	21	15	69
24·2	Maidenhead	..	..	..	25	26	25	72
31·0	Twyford	..	..	..	30½	32	10	70½
36·0	**READING**	..	..	..	36	36	30	43 (slack)
37·8	Southcote Jc.	..	..	..		39	00	
41·3	Theale	..	..	..		42	30	65
46·8	Midgham	..	..	..		47	35	62½
53·1	**NEWBURY**	..	..	..	54	53	30	
58·5	Kintbury	..	..	..		58	40	63½
61·5	Hungerford	..	..	..		61	35	
66·4	Bedwyn	..	..	..	67	66	20	61
70·1	Savernake	..	..	..		70	10	50
75·3	Pewsey	..	..	..		75	00	
81·1	Patney	..	..	..		79	40	
86·9	Lavington	..	..	..		84	15	82
91·4	Edington	..	..	..		87	40	
95·6	**WESTBURY**	..	..	..	94	91	30	30 (slack)
101·3	**FROME**	..	..	..		98	00	
106·6	Witham	..	..	..		103	35	56
108·5	Brewham Summit		..	..		105	45	51
115·4	Castle Cary	..	..	..	116	111	20	82/65
122·4	Charlton Mackrell	..	..	..		117	05	
125·7	Somerton	..	..	..		120	00	
131·0	Curry Rivel Jc.	..	..	..		124	40	76
						p.w. slack		
137·9	Cogload Jc.	..	..	..	137	131	30	
						p.w. slack		
142·9	**TAUNTON**	..	..	..	141½	141	00	
144·9	Norton Fitzwarren	..	..	..		144	50	
150·0	Wellington	..	..	..		150	40	52½
	Milepost 173	..	..	..				39½
153·8	Whiteball Box	..	..	..		155	48	
158·8	Tiverton Jc.	..	..	..		160	35	
165·3	Hele ..	..	..	..		165	50	79 (max.)
170·3	Stoke Canon	..	..	..		170	00	
173·7	**EXETER**	..	..	..	171½	172	55	63
178·5	Exminster	..	..	..		177	10	73
182·2	Starcross	..	..	..		180	30	
188·7	Teignmouth	..	..	..		187	20	
193·8	**NEWTON ABBOT**		..	..	195½	193	25	
	Net time—186½ minutes.							

*Engine later renamed *King George VI.*

booked passing time. In 2 min. 40 sec. the Limited was under way again, with a Duke class 4-4-0 piloting No. 6028, and with further excellent running Plymouth was reached in precisely four hours from Paddington.

Although the maximum load for a King over the South Devon line was fixed at 360 tons, when things were going well drivers were not particular about taking a few extra tons. In 1935, within a few days of each other, I recorded two fine examples of King performance west of Exeter. On the down journey, with 11 coaches, 376 tons tare, 405 tons full, the driver of No. 6016 *King Edward V* took no assistance with this heavy train, and made the run from Exeter to Plymouth, 52 miles, in 69 min. 25 sec. start to stop. Coming up, engine No. 6007 *King William III* had exactly 360 tons tare, on the up Cornish Riviera Express, and passed Newton Abbot comfortably in 45 min. 22 sec. On my down run with No. 6016 the time from Newton to Plymouth was 45 min. 43 sec. despite a heavy slack in the approach to North Road station. But on two other recorded occasions when the gross load was again 400 tons, or slightly over, first No. 6026 covered the Newton-Plymouth stretch in 42 min. 20 sec., and then No. 6016 achieved a time of 42 min. 45 sec.—very little slower, with 11-coach trains, than the $41\frac{1}{2}$ minute timing of the winter non-stop run, when the load was usually seven coaches, or at most eight.

Such work as this is enough to show the capacity of the Kings for continuous hard slogging, and a further example, if one were needed, is to be seen in a run on which engine No. 6022 *King Edward III* worked a load of 16 coaches, 537 tons tare, 575 tons full, throughout from Exeter to Paddington. This run was made after the construction of the Frome and Westbury avoiding lines, and the distance was thereby slightly reduced to 173·5 miles; but to work this vast load to Paddington in 177 min. 10 sec. despite two slight checks, was certainly one of the highest of "highlights" in this chronicle of the Kings.

This run was published by Mr. Cecil J. Allen in the *Railway Magazine* for September, 1933, and its details may be studied from the log overleaf.

G.W.R. EXETER—PADDINGTON

Engine—6022 *King Edward III*
Load—16 cars, 537 tons tare   575 tons full

Dist. miles					Schedule min.	Actual min. sec.		Speeds m.p.h.
0·0	EXETER	..	..	..	0	0	00	
3·5	Stoke Canon	..	..	..		6	15	47
7·2	Silverton	..	..	..		11	00	51½
12·6	Cullompton	..	..	..		17	00	55½
14·9	Tiverton Jc.	..	..	..		19	45	45
19·9	Whiteball Box	..	..	..		26	30	36
23·7	Wellington	..	..	..		30	15	82
28·8	Norton Fitzwarren	..	..	..		34	05	
30·8	TAUNTON	..	..	..	33	35	55	57½ (slack)
35·8	Cogload Jc.	..	..	..	38	40	10	
42·7	Curry Rivel Jc.	..	..	..		46	55	70
48·0	Somerton	..	..	..		52	15	54/69
58·4	Castle Cary	..	..	..	61	62	00	64
65·2	Brewham Summit		..	..		69	55	39
71·2	Blatchbridge Jc.	..	..	..		76	15	69
						p.w. slacks		
78·9	Heywood Rd. Jc.	..	..	..		86	50	40 (slack)
86·6	Lavington	..	..	..		95	45	64
92·4	Patney	..	..	..		101	55	53
98·2	Pewsey	..	..	..		107	35	65
103·4	Savernake	..	..	..		112	50	57
107·1	Bedwyn	..	..	..	112	116	25	
112·0	Hungerford	..	..	..		120	35	71
120·4	NEWBURY	..	..	..	124	127	30	75
126·8	Midgham	..	..	..		132	50	
135·6	Southcote Jc.	..	..	..		140	35	
137·5	READING	..	..	..	142	143	05	
142·5	Twyford	..	..	..	147	148	35	
149·3	Maidenhead	..	..	..	153½	154	40	74
155·0	SLOUGH	..	..	..	159	159	10	76½
160·3	West Drayton	..	..	..		163	30	
164·2	Southall	..	..	..	168	167	05	
167·8	Ealing	..	..	..		170	00	68
172·2	Westbourne Park	..	..	..		174	10	
173·5	PADDINGTON	..	..	..	179	177	10	

By this time the second batch of Kings was at work; these were named as follows:

6020	King Henry IV
6021	King Richard II
6022	King Edward III
6023	King Edward II
6024	King Edward I
6025	King Henry III
6026	King John
6027	King Richard I
6028	King Henry II
6029	King Stephen

The Kings

Last of the line, No 6029 *King Stephen* as originally built.
Later renamed *King Edward VIII*   [*British Rail*

Experimental streamling applied to No 6014 *King Henry VII*
as modified in 1935. Though the casing was later removed,
the locomotive kept its U-fronted cab to the end of its life
                                                [*British Rail!*

The final form: No 6011 *King James I* with double blastpipe
and chimney   [*G. W. Morrison*

Engine No 6005 *King George II* on Birkenhead to Paddington express leaving Harbury Tunnel, south of Leamington Spa
[*H. Weston*

Engine No 6018 *King Henry VI* on the down "Cornish Riviera Express" near Dawlish
[*British Rail*

Engine No 6007 *King William III* on the down "Cornish Riviera Express" climbing Wellington Bank in later Great Western days
[*F. R. Hebron*

King Edward II's driver peers into Dainton Tunnel's gloom as the locomotive heads the up "Cornish Riviera Express" on January 1, 1954
[*D. J. Fish*

Engine No 6005 *King George II* at Swindon in 1932 fitted with indicator shelters for dynamometer car test runs

[*P. Ransome Wallis*

Kings on Test

The Interchange Trials of 1948. Engine No 6018 *King Henry VI* ready to leave Kings Cross on a Leeds express

[*C. C. B. Herbert*

One of the finest runs I have ever seen on the Birmingham service was recorded by my friend Mr. A. V. Goodyear, on the 2.10 p.m. down, when the full load was 490 tons, as far as Banbury; was 435 tons to Leamington, and 400 tons on to Birmingham. The engine concerned was No. 6008 *King James II*. The log of this grand piece of running is included herewith. At Aynho Junction the train was just comfortably ahead of time; but a signal stop in the

2.10 p.m. PADDINGTON—BIRMINGHAM

Engine—6008 *King James II*

Load to Banbury	457 tons tare	490 tons full
to Leamington	405 tons tare	435 tons full
to Birmingham	370 tons tare	400 tons full

Dist. Miles					Schedule min.	Actual min. sec.		Speed m.p.h.
0·0	PADDINGTON	..	..	..	0	0	00	
3·3	Old Oak Common W.	..	..	7	7	33	40
7·8	Greenford	..	..	..		13	03	58
10·3	Northolt Jc. ..	..	..	..	15½	15	37	58
14·8	Denham	..	..	..	..		20	05	63
17·4	Gerrards Cross	..	..	..		22	45	54½
21·7	Beaconsfield	..	..	..		27	27	59
	Tyler's Green	..	..	..				69
26·5	HIGH WYCOMBE	..	..	..	32	31	50	42 (slack)
28·8	West Wycombe	..	..	..		34	45	
31·5	Saunderton	..	..	..		37	58	50
34·7	PRINCES RISBOROUGH		..	..	41	41	18	70
40·1	Haddenham	..	..	..		45	27	83½
44·1	Ashendon Jc.	..	..	..	49	48	36	52 (slack)
47·4	Brill	..	..	..	..		52	02	69
50·4	Blackthorn	..	..	..	..		54	30	75
53·4	BICESTER	..	..	..	..	58	56	55	67½
57·2	Ardley	..	..	..	..		60	32	59½
								68
62·4	Aynho Jc.	..	..	..	..	67	65	20	60
67·0	Milepost 85¾	..	..	..		70	27	sig. stop
67·5	BANBURY ..	..	..	..	pass 72	77	00	
0·0						0	00	
3·6	Cropredy	..	..	..	..		6	20	51
8·7	Fenny Compton	..	..	..		11	35	73½
13·7	Southam Road	..	..	..		15	40	73
15·9	Fosse Road Box	..	..	..		17	30	81
19·8	LEAMINGTON	..	..	..	19*	21	28	
0·0					0	0	00	
2·0	WARWICK ..	..	..	..		3	20	54½
6·2	Hatton	..	..	..	..		8	12	46
10·4	Lapworth	..	..	..	..		12	24	64
12·9	Knowle	..	..	..	..		14	51	63
16·3	Solihull	..	..	..	..		17	57	68
20·1	Tyseley	..	..	..	..		20	58	75
22·2	Bordesley	..	..	..	..		22	40	
23·3	BIRMINGHAM	..	..	..	26	24	35	

* Booked time from passing Banbury

C

approach to Banbury, and the slow resulting made it necessary to stop a second time to detach the Banbury slip portion. But some very fast work was done between Leamington and Birmingham; the time for this 23·3 miles was no more than 24 min. 35 sec., and included a sustained minimum speed of 46 m.p.h. on the 1 in 105 gradient of Hatton Bank. This latter feat probably entailed an output of more than 2000 indicated horsepower. In any case these runs are between them sufficient to show that the boiler and firebox of the Kings could amply sustain the efforts of a 40,000 lb. tractive-effort engine. If no other proof were necessary, the run of No. 6022, averaging 59 m.p.h. from Exeter to Paddington with the 16-coach train of 575 tons, would provide it.

CHAPTER SIX

A World Record for *Tregenna Castle*

IN THE SUMMER of 1929 the Castles came once again into the limelight. Since the construction of the Kings and the wonderful weight-hauling achievements of the latter engines, Mr. Collett's smaller four-cylinder 4-6-0s had naturally suffered a slight eclipse. But when the 2.30 p.m. express from Cheltenham to Paddington was accelerated to make the 77·3 mile run up from Swindon in 70 minutes, instead of the previous 75—a start-to-stop average of 66·3 m.p.h.— the capacity of the Castles for high sustained speed with moderate train-loads was manifested in most thrilling style. In the ordinary way the schedule was not unduly difficult. Of the intermediate timings 40 minutes were allowed for the 47·4 miles from Steventon to Southall, an average of 71 m.p.h.; and since the Castle class engines had shown their ability to run at 70 m.p.h. on level track with 500-ton trains, the working of the "Cheltenham Flyer," with its load of about 250 tons was no very severe tax. On the inaugural day of the accelerated schedule the famous engine of the 1926 trials on the L.M.S.R. was used, No. 5000 *Launceston Castle*, and with a load of 280 tons she completed the run in exactly 68 minutes despite one marked easing of the engine, and a slight signal check at Acton. Seventy miles in succession were run at an average speed of 73 m.p.h.

This was quite a model performance—no spectacular racing ahead of time, but just the minute or so in hand all the way, which is the mark of good driving. It was not long, however, before some considerably more exciting runs had been recorded. There were times when slacks for relaying were encountered; on occasions the train was a few minutes late in leaving Swindon; but from whatever cause, drivers seemed to revel in making up lost time, and then we had a taste of what the Castles could really do in the way of speeding on level track. One day when I was a passenger, No. 5003 *Lulworth Castle* was the engine, with a load of 275 tons; we attained 60 m.p.h. in 3½ miles from Swindon, but did not accelerate very rapidly afterwards until Uffington had been passed. To this point, 10·8 miles, the time was 11 mins. 55 secs., the speed was 73½ m.p.h. But then we really got going!

From Steventon to Reading the distance of 20·5 miles was covered in 14 min. 35 sec. at an average speed of 84·5 m.p.h., and twice in this distance, while running on level track, the speed reached 86½ m.p.h. By Reading the two minutes of lateness with which we had left Swindon had been recovered, and we continued at more moderate speed towards London. Even so the 26·9 miles from Reading to Southall were covered in 21 min. 23 sec.—an average of 74·7 m.p.h. Twice in the last few miles we were delayed by signals, but the run from Swindon was completed in 67 min. 15 sec. The net time on this run was no more than 64¼ minutes, equivalent to a start-to-stop average of 72 m.p.h. instead of the 66·3 m.p.h. scheduled. But it was, of course, that astounding spell from Steventon to Reading, at an average of 84·5 m.p.h. for 20 miles on end, that left so deep an impression. That was in 1929, and I was to remember that run many times in after years, for it proved verily the shape of things to come on the universally renowned "Cheltenham Flyer."

The next milestone in Great Western speed history came in the autumn of 1931, when three minutes were cut from the schedule of the "Cheltenham Flyer" and the start-to-stop average was raised from 66·3 to 69·2 m.p.h. This acceleration took place from September 14th of that year, and it was signalised by some remarkable running. Once again engine No. 5000 was chosen for the inaugural trip, but the load was somewhat lighter than in former years— six coaches totalling 190 tons behind the tender. Although the start was very fast, with a speed of 80·3 m.p.h. attained in less than six miles from Swindon, the standard of locomotive performance did not rise above that displayed on my own trip in 1929, with *Lulworth Castle*. But this time the speed was sustained without the slightest intermission from Shrivenham to Westbourne Park; this distance of 70·3 miles was covered in 50 min. 25 sec.—*fifty minutes, twenty-five seconds!*—an average speed of 83·6 m.p.h.

The running was kept at a remarkably even figure throughout, varying between 87 at Challow, 80 at Cholsey, 84 at Tilehurst, 80½ in Sonning cutting, and 85½ from Maidenhead to Slough. But the finish was exceptionally fast, and from Ealing to Old Oak Common the speed was sustained at 89 m.p.h. Thus the complete run from Swindon to Paddington was made in 59 min. 36 sec., an average of 77·8 m.p.h.

On the following day, again with *Launceston Castle*, according to the guard's journal an even faster run was claimed, but on the third trip on the accelerated schedule when Mr. C. J. Mount was a passenger, a world record was set up. With *Launceston Castle*, and a load of 195 tons, the run was actually made in 58 min. 20 sec, start-to-stop, giving an average speed of 79·5 m.p.h. This beat. by 1·2 m.p.h., the previous world record for steam traction made as long ago as 1905 on the Philadelphia and Reading Railroad when the run of 55½ miles from Camden to Atlantic City was made in 42 min. 33 sec. On the run of September 16th, 1931, *Launceston Castle* reached 90 m.p.h. on the very slight descent from Shrivenham to Steventon, and 67 miles were run at an average of 85·8 m.p.h. After thus demonstrating what Great Western engines could do, the Old Oak drivers were instructed that from September 17th onwards, no more "exhibition" runs were to be made. The scheduled time of 67 minutes was, of course, relatively easy of achievement. It is, however, significant of Castle capacity for continuous service that *Launceston Castle* was used day in day out for some time, and that in the first table of runs published in the *Railway Magazine* by Mr. Cecil J. Allen, a series of five, in December, 1931, no other engine appeared.

The Great Western Railway made the very most of the fact that on booked time alone the "Cheltenham Flyer" was the world's fastest train. A nameboard of striking design was carried on the front of the engine; souvenir luggage labels were designed, and Mr. H. M. Bateman drew a delightfully humorous cartoon to show what happened to a man who dared to pull the communication cord of the "Cheltenham Flyer." But the carrying of that special nameboard on the front of the engine must have perplexed many other people. The working of the "Cheltenham Flyer" formed part of a round trip from Paddington to Gloucester and back, and the outward run was made on the 10.45 a.m. from Paddington. Now if there was any Great Western long distance passenger train that could safely claim *not* to be the world's fastest train, it was the 10.45 a.m. of those days. But the turn-round time at Gloucester was only 57 minutes, and Old Oak Common sent out their Castle class engines fully prepared for the fast return journey; and this preparedness even included the "World's Fastest Train" board. Some passengers by the 10.45 a.m. who were not in "the know" may well have taken a curious view of railway speed!

But the Great Western authorities were not, even yet, satisfied with their achievements on the "Cheltenham Flyer," and on Monday, June 5th, 1932, they staged a speed exhibition in which not only the "Cheltenham Flyer" itself was concerned, but the corresponding down express leaving Paddington at 5 p.m., and the 5.15 p.m. up 2-hour Bristol express were also featured. So that the achievements of the locomotives should be corroborated beyond doubt the timing was done jointly by Mr. Humphrey Baker

THE CHELTENHAM FLYER

Date						June 6th, 1932		June 30th, 1937	
Engine No.						5006		5039	
Engine Name						*Tregenna Castle*		*Rhuddlan Castle*	
Load tons E/F						186/195		223/235	
Driver						Ruddock		Street	
Dist. miles				Schedule min.	Actual min. sec.	Speeds m.p.h.	Actual min. sec.	Speeds m.p.h.	
0·0	SWINDON	..	..	0	0 00		0 00		
5·7	Shrivenham	..	..		6 15	81½	6 35	80½	
10·8	Uffington	..	..		9 51	85½	10 10	90	
13·4	Challow	..	..		11 42	87	11 57	90	
16·9	Wantage Rd.		..		14 05	89½	14 12	94	
20·8	Steventon	..	..	18½	16 40	90	16 45	95	
24·2	DIDCOT	..	..	21	18 55	91½	18 57	90½	
28·8	Cholsey	..	..		21 59	91½	21 58	91	
32·6	Goring	..	..		24 25	92	24 25	88	
35·8	Pangbourne		..		26 33	90	26 40	86½	
38·7	Tilehurst	..	..		28 28	92	28 40	90	
41·3	READING	..	..	34	30 11	91	30 27	86½	
46·3	Twyford	..	..		33 31	89	33 50	90	
53·1	Maidenhead		..		38 08	87	38 17	93	
58·8	SLOUGH	..	..	47	42 10	86	42 08	90½	
61·1	Langley	..	..		43 42	86	43 38	86	
64·1	West Drayton		..		45 51	84	45 47	84	
68·2	Southall	..	..	54½	48 51	81½	48 46	82	
71·6	Ealing	..	..		51 17	84½	51 40	(eased)	
76·0	Westbourne Pk.		..	61	54 40		57 40		
77·3	PADDINGTON		..	65*	56 47		61 07		

*Schedule 67 minutes in June, 1932.

and Mr. Cecil J. Allen, and in actual fact the figures were further confirmed by various other observers who travelled in the trains concerned. The programme appeared to involve some pretty close timing on the part of the traffic department. The up express was then booked to leave Swindon at 3.48 p.m. and to reach Paddington at 4.55 p.m.; then the recorders were to take the 5 p.m. down express, normally non-stop to Kemble, but stopping at Swindon on this special occasion. Finally the 5.15 p.m. up 2-hour Bristol express was to be stopped specially at Swindon to pick up the recorders and to convey them back to London. This latter train was due to pass Swindon at 6.2 p.m. (only 62 minutes after the down Cheltenham express left Paddington) and was booked into Paddington at 7.15 p.m. This was close timing with a vengeance, if all was to go according to plan. In actual fact the up "Flyer" reached Paddington a shade before 4.45 p.m.; the down express reached Swindon at 6 o'clock almost to the second; the up Bristol express arrived at Swindon at 6.5 p.m. and left at 6.6, and the final arrival in Paddington was at 7.12½ p.m.

These mere times of the clock indicate some pretty startling running, which the detailed particulars given in the accompanying logs amply confirm. Engine No. 5006 *Tregenna Castle* was on the

5 p.m. Ex-PADDINGTON—JUNE 6th, 1932

Engine—5005 *Manorbier Castle*
Load—199 tons tare   210 tons full
Driver—Burgess (Old Oak Common)

Dist. miles					Schedule min.	Actual min. sec.		Speeds m.p.h.
0·0	PADDINGTON	..	..	..	0	0	00	
1·3	Westbourne Park	..	..	..		2	34	
3·0	Milepost 3	..	..	..		4	31	63
5·7	Ealing	..	..	..	..		6	52	74½
9·1	Southall	..	..	..	..	11	9	25	81
13·2	West Drayton	..	..	..	..		12	26	85½
18·5	SLOUGH	..	..	..	..	20	16	03	86½
24·2	Maidenhead	..	..	..	..	25½	20	14	82½
31·0	Twyford	..	..	..	..	31½	25	17	79
36·0	READING..	..	..	..	..	37	29	01	81½
38·6	Tilehurst	..	..	..	..		30	57	83
41·5	Pangbourne	..	..	..	..		33	03	83
44·7	Goring	..	..	..	..		35	24	83
48·5	Cholsey	..	..	..	..		38	03	84
53·1	DIDCOT	..	..	..	..	53	41	23	85½
56·5	Steventon	..	..	..	..	56½	43	47	83½
60·4	Wantage Road	..	..	..	..		46	35	81½
63·9	Challow	..	..	..	..		49	13	80
66·5	Uffington	..	..	..	..		51	13	78½
71·6	Shrivenham	..	..	..	..		55	03	81½
76·0	Milepost 76	..	..	..	..		58	15	83½
77·3	SWINDON	..	..	..	..	77	60	01	

up run, with a gross load of 195 tons. On that memorable afternoon the world record of 58 min. 20 sec. set up on September
16th, 1931, by engine No. 5000 was cut to 56 min. 47 sec., a truly
record average of 81·68 m.p.h. This was in the eastbound direction
where the gradient is definitely, if very slightly, in favour of the
engine. Swindon station is actually 270 ft. in altitude above the
level of Paddington, so that the average inclination is 1 in 1510.
Then on the return trip, and working against this slight gradient,
engine No. 5005 *Manorbier Castle* had a load of 210 tons. She
also put up a magnificent show, and reached Swindon in 60 min.
1 sec. As will be seen from the log the train was running at 80
m.p.h. just before Southall, and apart from two very brief pauses,
in Sonning cutting and near Uffington, the speed was maintained
at over 80 m.p.h. all the way to Swindon. The average over the
66·9 miles from Southall to Milepost 76 was 82·2 m.p.h. On the
up run *Tregenna Castle* averaged 90 m.p.h. for 39 miles on end!
Nor must the final achievement of the day be left unmentioned,
for on the up Bristol express No. 4091 *Dudley Castle* ran the 77·3
miles from Swindon to Paddington in 66 min. 33 sec.—equal to
the normal "Cheltenham Flyer" timing, inclusive of a signal check
to 55 m.p.h. approaching Didcot, and the usual slack to detach the
slip portion in the platform road at Reading.

Thus *Tregenna Castle* set up a new world record for steam
traction; what was more, the Castle class engines in general maintained a wonderful record of punctuality, even after the schedule
of the "Cheltenham Flyer" had been still further cut to 65 minutes.
That the great run of *Tregenna Castle* in 1932 did not represent
the limit of Castle achievement was shown some five years later
when a famous Old Oak driver, F. W. Street, made some even
faster running with engine No. 5039 *Rhuddlan Castle*, and a load of
235 tons. I have tabulated this journey alongside the world record
run of *Tregenna Castle*, and it will be seen that after a start of 20
sec. slower to Shrivenham *Rhuddlan* gradually caught up with
Tregenna, and had overhauled her by Cholsey. Then the two trains
ran "neck and neck" as it were, for the next seven miles, with *Tregenna
Castle* drawing away between Reading and Taplow. Then *Rhuddlan*
came up again in tremendous style with a maximum of 93 m.p.h.
near Taplow, against *Tregenna's* 87, and from Slough to Southall
the 1937 run took the lead. But here the train was so well ahead of
time that steam was shut off altogether, and 5039 ran into Paddington four minutes early. The most remarkable feature of this run was
the maximum speed of 95 m.p.h. at Steventon.

CHAPTER SEVEN

The Standard G.W.R. Express Passenger Engine

AT THE TIME of the great record runs of June, 1932, the Castle class was 46 strong, as against 68 of the Stars and 30 of the Kings. In addition to the regular series 4073-4099 and 5000-5012 there were five engines rebuilt from the Star class, including the famous 4000 *North Star* and No. 111. But from 1932 onwards no fewer than 85 new Castles were added to the stock, up to mid-summer of 1939. These engines were of unchanged design from No. 4073, though very slight and superficial external alterations were made, such as the shape of the front-end casing for the inside cylinders from engine No. 5013 onwards, and a shorter chimney from No. 5043 onwards. Of these additions Nos. 5083 to 5092 were rebuilt from the Abbey series of Stars. The successive batches were: 10 each in 1932, 1934 and 1935; 15 in 1936; 10 each in 1937, 1938 and 1939; and it was in 1938-9 at various times that the Stars 4063-4072 were renewed as Castles 5083-5092. In this another record fell to the Castles, for construction of no other British express passenger locomotive type proceeded to an unchanged design over so long a period as 16 years—1923-1939.

By 1938-9 the Castles were the standard express passenger class for the whole line. They could be used on all the express routes without restriction; they took over the heaviest duties in Cornwall, they went to Salisbury, Fishguard, Chester, and they handled the long and arduous through workings between Newton Abbot and Shrewsbury, and Newton Abbot and Wolverhampton. They were always available to deputise for the Kings on the heaviest West of England and Paddington-Wolverhampton turns, and on the Bristol main line they soon took over the working of the high speed "Bristolian" express, which ran the 118·3 miles from Paddington in 105 minutes, going down *via* Bath, and made the same time on the 4.30 p.m. up over the 117·6 miles from Bristol to Paddington *via* the Badminton route.

In the years just before the outbreak of the Second World War there were expresses making long start-to-stop runs at average speeds between 55 and 65 m.p.h. in every direction from Paddington and generally the standard of service was based on the everyday performance of the Castle class engines, as in later years. Not the least remarkable feature of the record runs with the Cheltenham Express in 1932 was the economical manner in which the engines themselves were driven. On the up journey *Tregenna Castle* was worked at between 17 and 18 per cent. cut-off with full regulator throughout from Shrivenham to Milepost 2, and on the down journey *Manorbier Castle* was worked at between 19 and 21 per cent. between Southall and Highworth Junction. From my own experience on the footplate I can say that 17 to 20 per cent. cut-off, with full, or nearly full, regulator represented the standard method of working Castle class engines in express passenger service.

Studying both pre-war schedules and those of 1949 in relation to the loads hauled it would seem that these were designed to be operated in standard Castle working conditions, leaving of course a little to spare for recovery from out-of-course slacks, and for the adding of extra coaches at week-ends, and so on. Of course, harder work was necessary in Cornwall, on the heavy grades between Bristol and Shrewsbury, and on the long ascent from Exeter to Whiteball tunnel, in maintaining the fast start-to-stop timing of 38 minutes for the 30·8 miles from Exeter to Taunton. But in general drivers aimed at getting back to 20 per cent. cut-off, or less, as soon as possible, so that the traditional working efficiency of the Castles could be realised. In carrying out tests on the stationary plant at Swindon Works it was usual to run locomotives at 20 per cent. cut-off and full regulator. In the recovery period since the end of the Second World War engineering restrictions, apart from any other circumstances, precluded the restoration of high speed running. A welcome measure of acceleration took place in May, 1946, and since then the timings have been generally such that the Castles could observe them with 14-coach trains, which the operating department regard as the normal maximum load.

Since that time, both while travelling as a passenger and when on the footplate, I have experienced a number of remarkably fine runs, showing a standard of performance no whit less than that of 1924 and 1932. The 7 a.m. from Weston-super-Mare to Paddington was a case in point. Before the war this train was booked to cover the 94 miles from Chippenham to Paddington in 89 minutes start-to-stop, an average speed of 63·4 m.p.h.; the usual load was about 270 tons, though this was reduced to about 200 tons by the detaching of a slip portion at Didcot. In 1949, with a working allowance of 104 minutes from Chippenham, the minimum load was 400 tons, with 365 tons from Didcot. But with considerably heavier loads than this Castle class engines have not merely maintained the present timing, but have even equalled the pre-war allowance. In 1939, 20 minutes were allowed to pass Swindon, 16·7 miles from the Chippenham start, and then 21 minutes for the 24·2 miles on to Didcot. On one splendid post-war trip, with engine 5056 *Earl of Powis*, and a load of no less than 470 tons, we passed Swindon in 20 min. 5 sec. and took 20 min. 55 sec. onwards to Didcot—exactly the pre-war timing! With a reduced load of 405 tons it would have been relatively easy to cover the remaining 53·1 miles to Paddington in 48 minutes but we were then before time, and went on quite easily. On a later trip No. 5009 *Shrewsbury Castle* made a still more brilliant start from Chippenham, with 455 tons, passing Swindon in 19 min. 20 sec.; and despite easier running afterwards we were still through Didcot in 41 min. 15 sec., practically the pre-war timing.

One of the most interesting journeys I have experienced on the footplate was with the 9.5 a.m. down, the equivalent in 1949 of the "Bristolian," then booked to cover the 118·3 miles in 2 hr. 18 min. with a stop at Reading. The normal load was one of 11 coaches, including a three-coach slip portion detached at Bath; and with this the allowance of 76 minutes from the Reading start to passing Bath (70·9 miles) was fairly easy. But my trip was made on a Monday morning, with two extra coaches and a gross load of 445 tons; this together with a permanent way slack near Maidenhead, and a bad signal check at Wantage Road, gave scope for some very fine running. No. 5029 *Nunney Castle* was the engine, and on her footplate I saw what could be done with the normal "full regulator—20 per cent. cut-off" style of working. Actually the cut-off was 19 per cent. from a point 1½ miles out of Paddington, and this gradually worked our 445-ton train up to 69 m.p.h. on the level at Slough. Equally fine work followed after the signal check at Wantage Road, where the driver linked up gradually to 19 per cent. again near Uffington. This was enough to take up us the gradual rise past Knighton Crossing to Shrivenham at 58 to 62 m.p.h., and by a characteristic piece of fast running west of Wootton Bassett, we were on time through Chippenham, and detached our Bath slip coaches on the stroke of 11.7 a.m. On the down gradients the regulator was eased back considerably, often using only the first valve, but the cut-off was not changed from 19 per cent. For 30 miles west of Shrivenham we averaged 67 m.p.h. with this heavy train.

Both Castles and Kings were most enjoyable engines to ride upon. Their action at the front end was very sweet, and even when they were putting forth a big output of power one rarely experienced any vibration, or sense of effort. On *Nunney Castle*, for example, after passing the foot of the Dauntsey incline with the engine travelling at 75 m.p.h., the driver opened out to practically full regulator; but apart from the exhaust becoming just audible there was no difference to be noted in the riding or in conditions on the footplate generally. During the war, and since its termination, both Castles and Kings have had to be fired often with coal far below pre-war standards. In my experience the steaming has always been excellent, though, of course, coal consuption rates have not always been at the low level recorded in the trials of *Caldicot Castle* in 1924. Some interesting tests were taken on the stationary plant in Swindon Works with different grades of coal; rather striking results were obtained when using the same grade of coal, though good sized lump in one case and three parts slack in the other. By analysis of the smokebox gases an accurate measure can be obtained of the amount of coal actually *burnt*, as distinct from the amount fed through the fire door. With Welsh coal, having a high calorific value but in a small and dusty condition it was shown that no less than 30 per cent. of that fired was going straight up the chimney unburnt!

Since the war, under the direction of Mr. F. W. Hawksworth, who succeeded Mr. Collett as Chief Mechanical Engineer in 1941, some interesting developments in the Castle design have been carried out; but before coming to these I should like to mention three further feats of performance by engines of the standard pre-war variety. The first of these was actually a pre-war run. In studying the details of the "Cheltenham Flyer" runs, when speeds up to 95 m.p.h. were made by the Castles on practically level road, one naturally asks what those engines might do if given their head on a stretch of favourable falling gradient. This question was to some extent answered when the 12.45 p.m. express from Paddington to Worcester achieved a maximum speed of 100 m.p.h. down the Honeybourne bank. But although the road itself is there very favourable, the distance in which to accelerate is very short, seeing that the descent is preceded by the rising stretch to Chipping Campden and the 1 in 100 gradient cannot be entered upon at more than about 65 m.p.h. Be that as it may, engine No. 4086 *Builth Castle*, hauling a load of 255 tons, accelerated from 65 to exactly 100 m.p.h. in four miles! Another startling exposition of how the Castles can accelerate was provided in the Automatic Train Control trial run of October, 1947. I was one of the privileged guests of the G.W.R. on that occasion, when No. 5056 *Earl of Powis* was working a four-coach test train of 142 tons gross. After passing Twyford at a little over 75 m.p.h., the engine was opened out and with a split-second chronograph I then recorded successive miles from Post 30 at 87·3, 90·5, 93·3, and 95·2 m.p.h., while the final half mile before brakes were applied for the test stop gave me a reading of 96·8 m.p.h. The exact maximum, as registered in the dynamometer car, was given as 96·4 m.p.h.

The last run to be mentioned was the first run I had seen since the war on which the Swindon-Paddington journey was made in less than even time. I have tabulated this journey as it is typical of post-war loading conditions, but no less typical of the ease with which good times could be made by the Castles when the road is clear. It is interesting also that this run was made by one of the very earliest Castles, now 25 years old. This was a fine example of the "standard" work of the class—steady, even running, without exceeding 69½ m.p.h. anywhere, and yet averaging 65·3 m.p.h. for

1.15 p.m. SWINDON—PADDINGTON—JULY 27th, 1948

Engine—4080 *Powderham Castle*
Load—13 cars—421 tons tare   455 tons full

Dist. miles						Schedule min.	Actual min. sec.		Speeds m.p.h.
0·0	SWINDON	..	..	..	..	0	0	00	
5·7	Shrivenham	..	..	..	..		8	52	62
10·8	Uffington ..	..	..	..	..		13	41	64½
13·4	Challow ..	..	..	..	..		16	10	66
16·9	Wantage Rd.	..	..	..	..		19	20	67
20·8	Steventon	..	..	..	..	22½	22	49	68½
24·2	DIDCOT	..	..	..	..	26	25	48	69
28·8	Cholsey ..	..	..	..	..		29	56	67
32·6	Goring ..	..	..	..	..		33	16	66
35·8	Pangbourne	..	..	..	..		36	13	66
38·7	Tilehurst ..	..	..	..	..		38	50	68
41·3	READING..	..	..	..	..	44	41	18	60 (eased)
46·3	Twyford ..	..	..	..	..	49	46	06	64½
53·1	Maidenhead	..	..	..	..	56	52	11	68½
58·8	SLOUGH	..	..	..	..	62	57	10	69½
61·1	Langley ..	..	..	..	..		59	09	67
64·1	West Drayton	..	..	..	..		61	54	66
68·2	Southall ..	..	..	..	..	72	65	43	64½
71·6	Ealing ..	..	..	..	..		68	48	67
75·3	Milepost 2	..	..	..	..		72	37	
76·0	Westbourne Park		..	..	..	81	73	35	
77·3	PADDINGTON		..	..	..	85	76	37	

70 miles on end. The result was an overall time of 76 min. 37 sec. from Swindon to Paddington, and on this July occasion of 1948 the load was not the 190 tons of the 1932 record but considerably more than double it.

CHAPTER EIGHT

Post-war Developments

IN 1941 Mr. Collett retired, and was succeeded as Chief Mechanical Engineer by Mr. F. W. Hawksworth. Like all the British railways, the Great Western was feeling acutely the strain of war conditions. Quite apart from the direct danger of enemy attack and dislocation of the traffic by bombing, there was the problem of coal. One could no longer depend upon getting the choice grades of Welsh soft coal on which so much of Great Western locomotive practice had been based in the past; moreover, there was always the possibility that with interruption of supplies the running sheds could not depend upon having the same grades from day to day, whether those grades happened to be good, bad or indifferent.

Towards the end of the war a series of experiments was commenced at Swindon to determine the degree of superheat giving the best all-round results. It was not only a question of economy; raising the degree of superheat increases the fluidity of the steam, and it flows more readily through ports and valves. There was thus a possibility of getting enhanced power from the locomotives through a more efficient use of the steam in the cylinders. And as the end of the war approached it became clear that the country itself was entering upon a period of unexampled austerity, and the duty rosters of locomotives would continue to be long and severe. Any changes in design that would help to keep engines on the road, at a time when attention would probably become less frequent was therefore worth while investigating.

The first trial of a higher degree of superheat was made on 10 Hall class 4-6-0s in 1944. This proved very successful, and after a further development, in the form of high-pressure County class 4-6-0s of 1945, the experiment was extended to the Castle class. In 1946 a new series was built, 5098-9 and 7000-7007, in which the Swindon superheater was replaced by a 21-element apparatus affording 295 sq. ft. of heating surface instead of the 262 sq. ft. provided by the standard 14-element Swindon superheater fitted to all previous engines of the Castle class. The number of small tubes was reduced from 201 to 170, and these changes resulted in a slight reduction of total heating surfaces, from 2,283 to 2,258 sq. ft. On account of the higher steam temperatures attained mechanical lubricators were fitted to supply the cylinders and piston valves, instead of the sight-feed lubricators fitted on earlier G. W. engines.

The new Castles proved to be very fast and powerful engines; by 1949 I had not had a footplate journey on one of them, but on one morning of very bad weather when I was a passenger on the 9.5 a.m. from Paddington, an engine of this class put up some very fast work west of Swindon, averaging 72 m.p.h. with a 455-ton load from Wootton Bassett to Corsham, and attaining a maximum of 82 m.p.h. on Dauntsey bank. Another excellent run was made by No. 7011 *Banbury Castle*, of the 1948 batch, on the 12 noon up express from Bristol, as between Swindon and Paddington. With a 13-coach train of 450 tons gross load behind the tender we made a splendid start, passing Shrivenham (5·7 miles) in 8 min. 18 sec. at 60 m.p.h. Then speed settled down to a steady 71-72 m.p.h. all down the fine stretch to Didcot, and this was followed by an average of 64·6 m.p.h. over the 48·8 miles from Didcot to Acton. So Didcot (24·2 miles) was passed in 24 min. 35 sec.; Reading (41·3 miles) in 39 min. 45 sec.; Slough (58·8 miles) in 56 minutes exactly, and a distance of 63 miles was covered in one hour from the dead start at Swindon. Acton (73 miles) was passed in 69 min. 41 sec., but the prospects of quite a record post-war trip were spoiled by a dead stand for signals at Portobello Junction, just on the country side of Westbourne Park. But the 75·8 miles from Swindon to this point were nevertheless covered in 72 min. 40 sec. start-to-stop, a fine average of 62·6 m.p.h. With a clear run in we could have stopped in Paddington in about 75¾ minutes from Swindon.

In the autumn of 1947, the experiments with higher degrees of superheat were carried a stage further, when a standard Castle engine, No. 5049 *Earl of Plymouth*, had the boiler re-tubed to include a four-row superheater, against the three-row apparatus fitted to the 5098 class. A locomotive inspector with whom I was riding about the same time on another Castle made the remark that No. 5049 was "all superheater", but again she appears to have done excellent work in service. At the end of 1947 she was put through some "full-dress" trials on the stationary testing plant at Swindon, for comparison with selected engines of the three-row series, and of old standard class with Swindon superheater. These trials with the different superheaters were extended, early in 1948, to include an engine of the King class, No. 6022 *King Edward III*, which was fitted with a four-row superheater. I was privileged to see this engine at "full speed" on the stationary test plant, and of all locomotive experiences I do not think I ever remember one more impressive, or more thrilling, up to that time.

At one time the capacity of this stationary testing plant was limited; but in later years the equipment was modernised and—

well, here was No. 6022 running at 60 m.p.h. with regulator full open and 20 per cent. cut-off, and developing between 1600 and 1700 indicated horsepower. A point that impressed me was the extraordinary quietness of the engine at speed; there was, it is true, a certain amount of noise, but it came rather from the brake wheels underneath. The exhaust was that lovely business-like purr, so characteristic of the Great Western four-cylinder 4-6-0s, and it was thrilling indeed to walk along the side and stand within a few inches of the revolving wheels and rods.

When the time came for the memorable locomotive exchanges of 1948 a standard King, No. 6018 *King Henry VI*, was chosen to make the running on the former L.N.E.R. line between King's Cross and Leeds. Just as Driver Young had done with *Pendennis Castle* in 1925, so Driver Russell impressed all onlookers by the way in which he lifted the 500-ton test trains away from King's Cross, without a trace of slipping. Owing to loading gauge restrictions King class engines could not be run on either the L.M.S.R. main line, or between Waterloo and Exeter. A short time after the exchanges were completed I rode from King's Cross to Grantham on an A4 Pacific with one of the drivers who had piloted Driver Russell during the trials, and he was full of praise for No. 6018.

The main series of trials were all carried out with the same grade of coal, Barnsley "hards," in all parts of the country; but later in the year some further dynamometer car trials were conducted in the Western Region to observe the difference in coal consumption when working the test trains on the best quality Welsh coal. After trials had been completed with a standard King class engine, the high superheat engine, No. 6022 *King Edward III*, was put through the same series of trials—1.30 p.m. Paddington to Plymouth, and 8.30 a.m. Plymouth to Paddington. On the last down trip the engine developed a minor mechanical defect just before leaving Old Oak Common, and so the test on the 1.30 p.m. had to be cancelled. But very soon the trouble was rectified, and so it was arranged to work the engine down on the 5.30 p.m. to Plymouth, and the opportunity was taken of conducting a full-dress trial on a much faster train than that featuring in the main series of exchanges during 1948. The normal load is one of 10 or 11 coaches, but on this occasion it was made up to 14, plus the dynamometer car— over 500 tons—with the full tonnage to be conveyed through to Exeter. So that the task set to No. 6022 can be better appreciated I have tabulated the working times of this train against those of the 1.30 p.m. of 1949 and those of the pre-war Cornish Riviera schedule, when a portion was slipped at Westbury.

Miles					Pre-War 10.30		1948 1.30		1948 5.30
0·0	Paddington	..	..		0		0		0
18·5	Slough	..	..		20		23		22
36·0	Reading	..	..	pass	36	{ arr. dep.	45 ‾‾ 50	pass	39
53·1	Newbury	..	..		54		72		57½
66·4	Bedwyn ..	..	..		67		87		72½
94·5	Heywood Rd. Jc.		..		—		117		99½
95·6	Westbury	..	..	pass	94	{ arr. dep.	120 ‾‾ 125		—
115·4	Castle Cary	..	..		116		149		121
142·9	Taunton	..	..		141½	{ arr. dep.	178 ‾‾ 184		148 ‾‾ 153
173·7	Exeter ..	..	..	pass	171½		222		189

I have it on good authority that No. 6022 put up a magnificent performance on the 5.30 p.m., which is only allowed six minutes more to Taunton than the fast pre-war schedule, although she took the full 500-ton load throughout.

Another post-war development, of which much was hoped but which faded out rather suddenly due to adverse economic conditions, was the change to oil-firing on some engines of the Castle class. Swindon made a first-class job of the conversions, and the engines themselves did some excellent work. Many Southern Railway engines were converted in a similar manner, and the policy of conversion was adopted and strongly advocated by the Ministry of Transport, until the "dollar situation" brought it to an end. Five Castles were converted:—

100	A. I Lloyds
5039	Rhuddlan Castle
5079	Lysander
5083	Bath Abbey
5091	Cleeve Abbey

I made some footplate journeys on 5039 and 5079 and found them excellent engines. *Lysander* was at that time stationed at Laira, and I rode on her in both directions between Plymouth and Penzance on the Cornish Riviera Express. As far as steaming was concerned one of her firemen aptly summed up the situation when he said: "You just can't knock her off the '225' "; and so I saw for myself in working the 365-ton up "Limited" on the long and severe ascent from Bodmin Road to Doublebois.

High speed automatic train control tests in 1947: engine
No 5056 *Earl of Powis* on a dynamometer car special near
Twyford at 86mph   [*M. W. Earley*

Castles on Trial

Oil firing in 1947: the up "Cornishman" passing Marazion
hauled by engine No 5079 *Lysander*

Dynamometer Car Tests

Engine No 6001 *King Edward VII* with improved draughting working a 25-coach controlled-test train, approaching Swindon at 60mph in 1954   [*British Rail*

The first double-chimneyed Castle, No 7018 *Drysllwyn Castle* on test with the up "Torbay Express" entering the loop at Churston, July 27, 1956   [*D. J. Fish*

Special Occasions

Funeral train of King George VI on February 15, 1952. The engine was actually No 7013, but was renamed and renumbered 4082 *Windsor Castle* for the occasion

[R. E. Vincent

The "Cornish Riviera Express" diverted via Swindon, here seen approaching Chippenham at 70mph. Engine No 6004 *King George III* *[Kenneth H. Leach*

Coronation Year, 1953. The "Inter-City" with special coronation headboard, climbing the bank out of Leamington Spa, with engine No 6013 *King Henry VIII* *[R. Blenkinsop*

Castle Chronology

The pioneer GWR 4cyl simple engine, *North Star* as rebuilt to the Castle class, running in BR days

[*Kenneth H. Leech*

The last Castle to be built, No 7037 named *Swindon* by Her Majesty the Queen, when she visited Swindon as Princess Elizabeth

[*British Rail*

The final version: engine No 5069 *Isambard Kingdom Brunel* with double blastpipe and chimney

[*F. D. Cassells*

CHAPTER NINE

Development under Nationalisation

AT THE TIME the first edition of this little book was completed, in 1949, construction of Castle class locomotives was still in progress at Swindon. These engines formed part of a programme authorised by British Railways to provide additional motive power pending the introduction of the new standard types, and so far as ex-Great Western designs were concerned included also batches of 4-6-0s of the Hall and Manor classes. So far as the Castles were concerned, the new engines were of the Hawksworth variety, with medium-degree superheat, mechanical lubrication, and straight-sided all-welded tenders. The running numbers eventually extended to No. 7037, and this last engine of the class was named *Swindon*, by Her Majesty The Queen, when as Princess Elizabeth she visited Swindon Works on November 15, 1950, and drove the Star class engine bearing her name.

The new engines, together with the earlier members of the class fulfilled their appointed functions admirably, in the difficult working conditions of the post-war years, setting up an excellent record of punctual running despite steadily worsening qualities of coal available, even for the highest class of express passenger duty. Generally, it was the occasional very bad supply that made conditions so difficult. With the very bad coal timekeeping was often out of the question, but it was the variation from day to day that most enginemen found so trying: reasonably good "Welsh" one day, and a mixture of kitchen nuts and ovoids the next. With the scheduled start-to-stop speeds of the principal express passenger trains in the 55 to 60 m.p.h. range, the train loads around 350 to 400 tons the overall results were satisfactory, and individual performance often as brilliant as at any time in the history of the G.W.R. four-cylinder 4-6-0 locomotives.

It was another matter altogether when the management of the Western Region decided to restore some of the fastest and most exacting pre-war schedules, including the "Bristolian", running each way between Paddington and Bristol in 1¾ hr., at average speeds of 67·7 m.p.h. down (via Bath) and of 67·2 m.p.h. up, over the slightly shorter route via Badminton. The programme included the restoration of pre-war speed with the Cornish Riviera Express, the Torbay Express, and the introduction of faster schedules than anything previously operated between Paddington, Newport and

Cardiff. Means had be to found of maintaining pre-war standards of reliability in steaming with the fuel available in the 1950s.

On the Great Western Railway, Churchward's locomotive practice had been based on the availability of ample supplies of first-class Welsh coal. Generations of enginemen had been trained to use it to the best advantage, maintaining boiler pressure at a few pounds per square inch below the rated blowing-off pressure throughout the run. This was no remote ideal. It was achieved day-in day-out in every express link on the line, and thus assured of constantly maintained boiler pressure Churchward could use a low degree of superheat, with advantageous results in the upkeep of boilers, tubes, valves and so on, and could use a device like the jumper ring on the blast pipe, which would automatically lift and reduce the back pressure, if a driver should tend to be heavy handed, and thrash his engine to the detriment of economic working. Every feature on the Star class engines was designed to promote low coal, water and oil consumption, and to keep maintenance charges to a minimum.

In the 1950s such almost Utopian conditions on the footplate could hardly be expected. The over-riding need was to have reliable steaming. A fractionally higher coal consumption was a secondary consideration. In these circumstances, the Western Region locomotive department was in the fortunate position not only of inheriting the former G.W.R. stationary testing plant, and dynamometer car, but also, through the farsightedness of F. W. Hawksworth, the last C.M.E. of the G.W.R., being well embarked on the development of entirely new testing techniques. After nationalisation these were adopted by the Railway Executive as the British standard method of testing, and accordingly the work at Swindon continued without intermission under the leadership of S. O. Ell.

On the stationary testing plant the technique of draughting was exhaustively studied, and as a result alterations were made to the smokebox layouts of both Kings and Castles. The jumper rings were removed, thus eliminating the upward limitation of draught previously imposed on all the locomotives of Churchward parentage, and by very careful design and experimental work a new combination of blastpipe and chimney arrived at which substantially increased the maximum steaming capacity of the engines. With first class coal the maximum steam rate was increased, in round terms, by about 20 per cent. over the former maximum. What was more important however was that with inferior coal something approaching pre-war steaming rates could be maintained. The basic coal consumption was increased, but by no more than a modest amount in relation to the increased reliability of the engines in the heaviest express traffic.

This "improved draughting", as it was termed, formed the second stage of the development from the original Churchward precepts. The first, initiated under Hawksworth, had been the use on both Castle and King class engines of a higher degree of superheat, combined with the use of mechanical lubricators. The results from improved draughting were so successful that the Locomotive Department could undertake, in every confidence, the working of the "Bristolian" at pre-war speed, as from the summer service of 1954. As when the service was first introduced, in 1935, King class engines were employed at first; but after a few months of running experience it was found possible to use Castles, without any deterioration in the standards of punctuality. The remarkable test runs with engine No. 6001 *King Edward VII* in 1953 which signalised the success of Ell's work at this stage, are referred to in the next chapter.

Before the acceleration of the Cornish Riviera Express to a four-hour non-stop run between Paddington and Plymouth, programmed for the summer service of 1955, some further test runs were carried out to convince the operating department of its practicability with King class engines in the prevailing conditions. By way of further demonstrations test runs were also made with a Stanier Pacific engine from the London Midland Region. The summer service of 1955 saw a more general acceleration of the express train services. including Torbay, and the promised improvements on the Paddington-South Wales runs.

In the early spring of 1955 the Swindon stationary testing plant was running the British Railways class '8' three-cylinder 4-6-2 engine No. 71000 *Duke of Gloucester*, which had a double blastpipe and chimney. At that time R. A. Smeddle was Chief Mechanical and Electrical Engineer, Western Region, and with a view to improving still further the performance of both the Kings and Castles he instructed Ell to prepare a double-exhaust layout for both classes. The improved draughting arrangement fitted to all the Kings and to many of the Castles had been a very simple and cheap conversion; but consideration had next to be given to a more thorough modernisation. The hard work entailed was showing up weaknesses in the frames of the Kings in particular—not surprisingly as the majority of the class was by then nearly 30 years old. Although plans for the replacement of steam by diesel traction were by then well under way it was evident that steam would have to carry on for a further term. No new engines of Class '8' capacity could be built, and so an extensive rebuilding programme was authorised.

Concurrently with the fitting of all the King class engines with double-blastpipes and chimneys, new cylinders were fitted, though cast to the original patterns. Using a technique first introduced at Crewe new sections of frame were fitted at the front end, cutting the old plates away, welding the new sections on to the rearward portions which were still serviceable. There is no doubt that the performance of the double-chimneyed engines would have been still further enhanced had the cylinders been re-designed to provide for full internal streamlining; but the cost of making the necessary new patterns precluded any such further refinement. A number of Castle class engines was also equipped with double blastpipes and chimneys, and although these showed an improved performance the difference was not considered to be such as to warrant altering the whole class. In any case many of the Castles were relatively new engines.

CHAPTER TEN

Classic Dynamometer Car Test Runs

FOLLOWING the Hawksworth developments in superheating a series of dynamometer test runs was made, between Paddington and Cardiff, with the three varieties of Castle originally provided, namely the Collett standard engine, with low superheat and hydrostatic lubricator; the Hawksworth medium superheat engine, and the experimental rebuilt engine No. 5049 *Earl of Plymouth*, with high degree superheating. Although mechanical lubrication was standardised on the medium superheat engines numbered from 7000 to 7037 the two original engines of this series, 5098 *Clifford Castle* and 5099 *Compton Castle* had hydrostatic lubricators, and it was No. 5098 that figured in the comparative trials. These trials were made on ordinary express trains at the moderate speeds then scheduled, and although the basic performance of all three engines was good there was nothing spectacular about the running.

It was the trials centred upon the fitting of the improved draughting to the Kings that provided some of the most impressive running data, and these trials fell into two groups. There was the basic tests on the stationary plant at Swindon, with engine No. 6001 *King Edward VII* to verify the results achieved, and to relate these to performance on the road in the form of a series of controlled road tests with special trains of empty stock between Reading and Stoke Gifford. Following these were certain special runs at service train timings to satisfy the still unconvinced traffic department that the Kings were capable of hauling trains at the accelerated times demanded. All the Swindon stationary tests and the associated controlled road tests were made with engine No. 6001 *King Edward VII*; the 'dress rehearsal' run for the Bristolian was made with No. 6003 *King George IV*, and the epoch-making runs with the Cornish Riviera Express in March 1955, in very rough weather, were made with No. 6013 *King Henry VIII*.

The tests on the Swindon stationary plant were conducted with good quality soft Welsh coal, to ascertain the maximum performance in ideal conditions for the re-draughted engine. The results achieved were contained in a most comprehensive paper read before the Institution of Locomotive Engineers in November 1953 by Mr. Ell, and some of the more spectacular results are tabulated herewith. Working at a constant steam rate of 33,600 lb. per hour, with a coal consumption of 5465 lb. per hour the following results were obtained.

Road speed m.p.h.	Cut-off %	Indicated Horsepower	Drawbar Horsepower
30	55	2050	1810
40	40	2150	1800
50	35	2160	1680
60	31	2160	1515
70	28	2150	1310
75	27	2140	1200

These figures show clearly how the internal resistances of the locomotive mount up with increased speed, so that whereas all but 240 horsepower of the 2050 developed in the cylinders is available for traction at 30 m.p.h. no less than 940 h.p. is absorbed internally in the locomotive at 75 m.p.h.

The maximum coal consumption that a single fireman could manage continuously was agreed at 3000 lb. per hour, and this provided the following power outputs at speeds between 30 and 75 m.p.h.

HORSEPOWER AT 3000 lb. of Coal per hour

Speed m.p.h.	Indicated Horsepower	Drawbar Horsepower
30	1550	1350
40	1670	1340
50	1740	1270
60	1770	1150
70	1780	970
75	1780	870

A last point to be noted in connection with these stationary plant trials is the extent the coal consumption increases when the boiler is pushed to its maximum output. A proportional ratio is given below:

Coal consumption lb. per hour	Steam rate lb. per hour
2850	24,000
3600	28,000
4065	30,000
5465	33,600

The tests carried out with special trains and the dynamometer car gave a most vivid impression of what these steam and coal rates involved. I was invited to travel on a test run when the steam rate was to be 30,000 lb. per hour, and this meant hauling a train of 25 coaches—a load of all but 800 tons. This enormous train was worked at standard express train speed, that is, the kind of speed one would expect on the South Wales service with a load of 400 to 450 tons.

To maintain steam pressure with a coal rate considerably above the 'one man' maximum two fireman were carried; but one man fired on the outward journey and the other fired on the return. The following is a log of the return trip.

STOKE GIFFORD—READING

Load: 796 tons tare   798 tons full
Engine: 6001 *King Edward VII*

Dist. miles							Time min. sec.		Speeds m.p.h.
0·0	*Stoke Gifford East*	..	..	..	..	..	0	00	—
6·9	Chipping Sodbury	..	..	..	..	..	13	58	45
11·5	Badminton	..	..	..	..	..	20	06	45½
21·3	Little Somerford	..	..	..	..	..	29	32	78
28·6	Wootton Bassett	..	..	..	..	..	35	43	58
34·2	SWINDON	..	..	..	..	..	41	34	60
45·0	Uffington	..	..	..	..	..	51	24	69
51·1	Wantage Road	..	..	..	..	..	56	38	70½
58·4	DIDCOT	..	..	..	..	..	62	48	71
70·0	Pangbourne	..	..	..	..	..	72	57	67
73·5	*Scours Lane Jc.*	..	..	..	..	..	76	55	—

The start up to Badminton is on a continuous incline of 1 in 300, and this is followed by the corresponding descent to Little Somerford where this huge train was travelling at 78 m.p.h. Then after the rise to Swindon came the long, very gradual descent to Didcot, and the final stretch to Pangbourne tapers off to dead level. It was certainly a wonderful performance, and remarkable in its complete verification of the predicted speeds after the stationary plant results had been analysed.

The 'dress rehearsal' for the Bristolian acceleration took place on April 30th 1954, and was a virtually continuous run of 235 miles, because the return journey, via Badminton was commenced only 19 min. after our arrival. Summary details of this remarkable round trip are set out in the logs printed on the following page. The load was 253 tons tare, 260 tons full, and the engine No. 6003 *King George IV*.

The round trip may be summarised by adding that we had covered 233·9 miles not including the short run from Dr. Day's Bridge Junction to Stapleton Road, in 194 min. 64 sec.—an average taking the two runs together of 72 m.p.h. This was considerably faster than what would normally be required on the Bristolian, and the accelerated schedule was operated from the inauguration of the summer service that year.

DOWN JOURNEY

Dist. miles							Time min.	sec.	Speeds m.p.h.
0·0	PADDINGTON	..	..	..	..	..	0	00	
9·1	Southall	..	..	..	..	..	10	18	72
18·5	Slough	..	..	..	..	..	17	51	82
36·0	READING	..	..	..	..	..	31	23	72
53·1	Didcot	..	..	..	..	..	44	44	84
66·5	Uffington	..	..	..	..	..	54	38	79½
—							p.w.s.		18
77·3	SWINDON	..	..	..	..	..	65	44	72
87·7	Dauntsey	..	..	..	..	..	73	20	96½
94·0	Chippenham	..	..	..	..	..	77	33	
106·9	BATH	..	..	..	..	..	88	17(slack)35	
113·8	Keynsham	..	..	..	..	..	94	57	75
117·9	Dr. Days Bridge Jc.	..	..	..	..	..	99	19	

Net time 96 min.

UP JOURNEY

Dist. miles							Time min.	sec.	Speeds m.p.h.
0·0	BRISTOL STAPLETON RD.			..	..	..	0	00	—
3·2	Filton Junction		..	..	..	..	5	10	—
11·4	Chipping Sodbury		..	..	..	..	13	30	66
16·0	Badminton	..	..	..	..	..	17	47	63½
26·3	Little Somerford	..	..	..	..	..	25	25	93
33·1	Wootton Bassett		..	..	..	..	30	22	65
38·7	SWINDON	..	..	..	..	..	35	03	79
49·5	Uffington ..	..	..	..	..	..	42	51	87
55·6	Wantage Road	..	..	..	..	..	46	59	90/88
62·9	DIDCOT ..	..	..	..	..	..	51	52	92½
—							eased		73
80·0	READING..	..	..	..	..	..	64	31	82
91·8	Maidenhead	..	..	..	..	..	73	01	90
97·5	Slough ..	..	..	..	..	..	77	08	81/86
106·9	Southall ..	..	..	..	..	..	84	13	—
—							p.w.s.		15
114·7	Westbourne Park	..	..	..	..	..	92	55	
116·0	PADDINGTON ..	..	..	..	..	..	95	35	

Net time 93½ min.

CHAPTER ELEVEN

Daily Performance in the Last Years

IN REVIEWING the work of the Kings and Castles in the last years before their superseding by diesels it is perhaps no more than natural to pick out some of the best runs from one's own experience. In my regular travelling on the Bristol main line, supplemented by

EXETER—PADDINGTON: TORBAY EXPRESS

Engine: 5059 *Earl St. Aldwyn*
Load: 11 cars   387 tons tare   425 tons full

Dist. miles						Sch. min.	Actual m.	s.	Speeds m.p.h.
0·0	**EXETER**	..	..	..	..	0	0	00	—
1·3	*Cowley Bridge Jc.*	..	..	..	..	3	3	12	
3·5	**Stoke Canon**	..	..	..	..		6	12	45½
8·4	**Hele**	..	..	..	..		12	10	53½
12·6	**Cullompton** ..	..	..	..	..		16	44	57
14·8	**Tiverton Junc.** ..	..	..	..	..		19	17	50/56
19·9	*Whiteball Box* ..	..	..	..	..	24½	25	27	40½
23·7	**Wellington** ..	..	..	..	..		29	00	82
28·8	**Norton Fitzwarren**	..	..	..	..		32	39	86
30·8	**TAUNTON** ..	..	..	..	..	32½	34	05	80
33·2	*Creech Junc.* ..	..	..	..	..	34	35	54	82
42·5	*Curry Rivel Junc...*	..	..	..	..		43	49	70/75
48·0	**Somerton**	..	..	..	..		48	29	60/69
—							sig.	stop	5¼ min.
58·4	**Castle Cary** ..	..	..	..	..	60	67	40	—
65·2	*Milepost 122¾* ..	..	..	..	..		79	51	29
71·2	*Blatchbridge Junc.*	..	..	..	..		86	12	75/69
76·5	*Fairwood Junc.* ..	..	..	..	..	77½	90	33	77½
86·6	**Lavington** ..	..	..	..	..		99	27	69
91·5	**Patney**	..	..	..	..	92	105	13	54½
98·2	**Pewsey**	..	..	..	..		111	00	64½
103·4	**Savernake** ..	..	..	..	..	102½	116	13	56
120·4	**NEWBURY** ..	..	..	..	..	121	131	00	76 (max)
135·6	*Southcote Junc.* ..	..	..	..	..		144	13	
							sigs.		
137·5	**READING..** ..	..	..	..	..	136½	148	12	
149·3	**Maidenhead** ..	..	..	..	..	147½	159	51	71½
155·0	**SLOUGH**	..	..	..	..	152	164	41	72
164·4	**Southall**	..	..	..	..	160	173	02	66/70½
172·2	**Westbourne Park**	..	..	..	..		179	43	
173·5	**PADDINGTON** ..	..	..	..	..	175	183	07	

Net time 167 min.

a number of special trips made specially for observation of the
locomotive work, I was fortunate in securing a number of very
fine records, though other observers, who travelled regularly were
no less fortunate. The runs that are described in this chapter, all
except two of my own recording, are typical rather than exceptional of the best work of the two famous classes.

The first, details of which are tabulated, was on the up Torbay
Express on its summer mid-week working, and scheduled to cover
the 173·5 miles from Exeter to Paddington in 175 min. On this
occasion I was on the footplate, with Driver Trendall and Fireman
Bascomb of Newton Abbot. The initial allowance of 34 min. to
pass Creech Junction is extremely sharp, including a point-to-point
time of 9½ min. for the 13·4 miles downhill from Whiteball summit.
The driver did not press his engine unduly up the long bank from
Exeter, and a fast descent of Wellington bank with a sustained
maximum speed of 86 m.p.h. was not enough to keep point-to-
point time on that stage. But there was method in this working
because time could be gained over the Langport cut-off line, and
the climbing of the long 1 in 264 rise to Somerton tunnel at a minimum of 60 m.p.h. was an excellent piece of work.

Then unfortunately a minor mishap to a preceeding train caused
a signal stop of 5½ min. before Castle Cary, and further delay
afterwards so that after climbing the Bruton bank, at much lower
speed than normal, we were 13¼ min. late on passing Blatchbridge
Junction. By that time we were going well again, and 5 min. were
recovered on the fast schedule onwards to Paddington, despite a
signal check outside Reading. This schedule was planned for the
working of a 350-ton load, and included an 8-min. recovery time.
Our running showed a net gain of exactly 8 min., but this had been
achieved with one extra coach, making a tare load of 387 tons.
Although this shows a standard of performance above what was
planned for the service this would not be regarded as anything more
than a typically good Castle run.

The work of the Castles at its very finest is shown by the second
run tabulated, on the 8.30 a.m. up from Plymouth. The engine was
the last but one built, No. 7036 *Taunton Castle*, worked by Driver
A. Cook and Fireman Hughes of Old Oak Common. Despite a
load of 73 tons heavier, which occasioned a slower start, the hill-
climbing from Stoke Canon to Whiteball was faster than that of
Earl St. Aldwyn on the Torbay Express, and a fast descent from
Whiteball ensured timekeeping on the first stage, despite a permanent-way check to 15 m.p.h. At Taunton yet another coach was
added to this already very heavy train, and the locomotive authori-

10.3 a.m. EXETER—WESTBURY

Engine: 7036 *Taunton Castle*

Load to Taunton   14 cars   460 tons tare   500 tons full
    to Westbury 15 cars   490 tons tare   535 tons full

Dist. miles							Sch. min.	Actual m.	s.	Speeds m.p.h.
0·0	EXETER	..	..	..	..	..	0	0	00	—
1·3	*Cowley Bridge Jc.*		..	..	..	..		3	41	—
3·5	Stoke Canon	..	..	..	..	..		6	45	48½
7·2	Silverton	..	..	..	..	..		11	05	55½/53
8·4	Hele	..	..	..	..	..		12	24	57
12·6	Cullompton	..	..	..	..	..		16	45	59
14·8	Tiverton Junc.	..	..	..	..	..		19	16	50½/57½
19·9	*Whiteball Box*	..	..	..	..	..	25	25	29	37½
23·7	Wellington	..	..	..	..	..		29	19	77½
26·9	*Milepost 167*	..	..	..	..	..		31	47	82
—								p.w.s.		15
28·8	Norton Fitzwarren		..	..	..	..		33	32	
30·8	TAUNTON	..	..	..	..	..	38	36	50	
2·4	*Creech Junc.*	..	..	..	..	..	4	4	30	51
4·7	Cogload	..	..	..	..	..		7	00	56
8·0	Athelney	..	..	..	..	..		10	18	62/59
11·9	*Curry Rivel Jc.*	..	..	..	..	..		14	17	61½
17·1	Somerton	..	..	..	..	..		19	54	53/63½
20·5	Charlton Mackrell		..	..	..	..		23	14	57
22·7	Keinton Mandeville		..	..	..	..		25	26	66½
27·5	Castle Cary	..	..	..	..	..	31	30	03	58/60
31·0	Bruton	..	..	..	..	..		33	44	50½/52½
33·1	*Milepost 124*	..	..	..	..	..		36	21	46½
34·4	„ 122¾	..	..	..	..	..		38	11	34½
40·4	*Blatchbridge Junc.*		..	..	..	..	46	44	16	72
42·4	*Clink Road Junc.*..		..	..	..	..	48½	46	01	64½
45·7	*Fairwood Junc.*	..	..	..	..	..	52½	49	01	72
47·2	WESTBURY	..	..	..	..	..	55	51	39	

ties had an assistant engine standing ready to couple on. But *Taunton Castle* was in such magnificent form that any suggestion of assistance was disdained, and with a load of 535 tons a magnificent run was made to Westbury. Outstanding features were the sustained minimum speed of 53 m.p.h. up the 1 in 264 gradient to Somerton tunnel, and the remarkable minimum speed of 34½ m.p.h. at the summit of the severe Bruton bank. With some fast downhill running in conclusion we not only kept time, but registered a *gain* of 3¼ min.

on schedule. Unfortunately I was travelling to Chippenham on this occasion, and so, leaving the train at Westbury, I was not able to record how this outstanding engine conveyed the train onwards to Paddington.

In contrast to these heavy-load runs on the West of England main line the working of the Castles on the high speed Bristolian express always provided some excitement. There are tabulated two runs on the up service with the double-chimneyed engine No. 7018 *Drysllwyn Castle*, which was always a very free runner. On the first of the two runs, with Driver W. H. Brown, I was on the foot-

BRISTOL—PADDINGTON: THE BRISTOLIAN

Engine: 7018 *Drysllwyn Castle* (double chimney)

Dist. miles		Sch. min.		Load tons E/F 247/265 Driver: W. H. Brown Actual m. s.		Load tons E/F 247/260 J. Russe Actual m. s.	
0·0	BRISTOL	0	..	0	00	0	00
1·6	Stapleton Rd. ..		..	4	08	4	00
4·8	Filton Jc.	8½	..	9	40	9	03
7·7	Winterbourne ..		..	12	54	12	35
13·0	Chipping Sodbury		..	18	00	17	25
17·6	Badminton ..	21½	..	22	11	21	15
27·9	Little Somerford		..	29	48	28	05
34·7	Wootton Bassett	34	..	34	43	32	48
40·3	SWINDON ..	39	..	39	26	37	29
46·1	Shrivenham ..		..	43	43	41	40
53·7	Challow		..	49	05	46	55
61·1	Steventon ..		..	53	58	51	55
64·5	DIDCOT	59½	..	56	13	54	15
72·9	Goring		..	61	49	59	57
78·9	Tilehurst		..	66	19	64	13
81·6	READING.. ..	71½	..	68	20	66	11
86·6	Twyford	75	..	72	12	70	00
93·4	Maidenhead ..	80	..	77	09	75	03
99·1	SLOUGH ..	84½	..	81	10	79	07
104·4	West Drayton ..		..	84	57	82	47
108·5	SOUTHALL ..	91	..	88	01	85	45
111·9	Ealing		..	90	33	88	11
116·3	Westbourne Park		..	94	28	91	45
117·6	PADDINGTON	105	..	97	08	93	50

Speeds (m.p.h.)

Min. at Filton	..	..	..	30		33	
„ at Badminton	..	..	..	67/66		73	
Max. at Little Somerford ..	..	..	..	94		100	
Average: Shrivenham to Goring		..	..	88·8		87·9	

plate. The start, up the 1 in 75 gradient of Filton incline, always needed careful handling, because if an engine was pounded hard to keep strict point-to-point time at Filton Junction, the fire could be torn about, and the resulting holes would cause poor steaming for many miles afterwards. Thus the engine was not pressed, falling to a minimum speed of 30 m.p.h., and she was allowed to find her own pace up the long 1 in 300 gradient to Badminton. By that time generally the engine of the up Bristolian was thoroughly warmed up, and producing a fine output of power.

On the corresponding 1 in 300 descent to Little Somerford a speed of 94 m.p.h. was attained, and then after easing through Wootton Bassett, and taking the slight rise to Swindon there came a long sustained fast spell of running. Speed averaged 88·8 m.p.h. over the 26·8 miles from Shrivenham to Goring, with a maximum of 91 m.p.h., and with fine subsequent running and a complete absence of checks Paddington was reached 8 min. early.

The companion run, logged by a friend who is a very experienced recorder, was made at a time when certain very hard runs were made to determine whether some further acceleration of schedule might be possible. I believe it is the fastest end-to-end time made by a steam locomotive in the up direction. The engine was pressed rather harder than usual in the early stages, with speeds of 33 m.p.h. up the Filton bank, and an acceleration to 73 m.p.h. at Badminton. A maximum speed of exactly 100 m.p.h. at Little Somerford put the finishing touch on a remarkable initial effort, by which Driver Russe gained two min. on Driver Brown by the time Shrivenham was passed. From there onwards to Ealing the two runs virtually dead-heated, with times of 46 min. 50 sec. and 46 min. 31 sec. over this stretch of 65·8 miles. After that Driver Russe made the faster finish, and ended with the record time of 93 min. 50 sec. from Bristol to Paddington.

In the early days of the restored 1¾-hour service between Paddington and Bristol King class engines were used; but it was soon found that the schedule could be satisfactorily managed by Castles. The next table nevertheless sets out details of two remarkable runs with Kings. The first, with engine No. 6018 *King Henry VI*, Driver Pithers and Fireman French, of Old Oak was no more than an ordinarily good run until after Swindon when an exceptionally fast descent was made of Dauntsey bank, with a maximum speed of 102½ m.p.h. This is the fastest I have personally logged with any Great Western engine, and it is the fastest fully-authenticated speed recorded with a King. There have been reports of somewhat higher speeds, including one of 110 m.p.h. by the first engine of the class to be fitted with a double-chimney No. 6015 *King Richard III*. This feat was claimed when two members of the staff of the Chief Mechanical Engineer were riding on the footplate; but without any details of point-to-point times it is not possible to accept the precise authenticity of this claim.

PADDINGTON—BRISTOL: THE BRISTOLIAN

Engine No. Engine Name Load tons E/F					6018 *King Henry VI* 236/250		6015* *King Richard II* 279/300	
Dist. miles				Sch. min.	Actual m.	s.	Speeds m.	s.
0·0	**PADDINGTON**	..	0	..	0	00	0	00
1·3	Westbourne Park	..		..	2	56	2	48
9·1	Southall ..	..	11	..	10	44	10	41
13·2	West Drayton	..		..	14	01	13	52
—					—		sig. stop	
18·5	**SLOUGH**	..	17½	..	17	57	20	22
24·2	Maidenhead	..		..	22	12	26	29
—					p.w.s.		—	
31·0	Twyford ..	..		..	27	29	31	30
36·0	READING..	..	31	..	31	54	34	57
41·5	Pangbourne	..		..	36	00	38	50
48·5	Cholsey ..	..		..	41	15	43	37
—					sigs.		sigs.	
53·1	DIDCOT ..	..	46	..	45	15	48	50
—					—		p.w.s.	
60·4	Wantage Road	..		..	51	47	59	57
66·5	Uffington ..	..		..	56	27	65	40
71·5	Shrivenham	..		..	60	09	69	36
—					p.w.s.		—	
77·3	**SWINDON**	..	67	..	65	31	73	59
82·9	Wootton Bassett	..		..	69	43	77	55
87·7	Dauntsey	..		..	72	53	81	01
94·0	**CHIPPENHAM**	..	79	..	77	00	85	03
98·3	Corsham	..		..	80	39	88	12
101·9	Box	..		..	83	23	90	44
106·9	**BATH** ..	..	90½	..	88	12	95	06
113·8	Keynsham..	..		..	95	16	101	03
116·7	St. Annes Park	..		..	98	32	103	15
—					—		sigs.	
118·4	BRISTOL ..	..	105	..	102	05	109	24

Net times		min.	99	92

Speeds (m.p.h.)

Net average, Southall—Swindon	79.2	83·1
Max. at Dauntsey	102½	98

* Fitted with double chimney.

This same engine was concerned in the very fast run on the down Bristolian set out in the table. The load was considerably above normal, added to which there were some unfortunate delays. Intermediately the engine was worked extremely hard, as will be evident from such a time as 23¼ min. over the 30 miles from Slough to Cholsey with its average speed of all but 80 m.p.h. The net average speed, from Southall to Swindon was indeed no less than

PADDINGTON—NEWTON ABBOT:
CORNISH RIVIERA EXPRESS

Engine: 6013 *King Henry VIII*

Load to Heywood Rd.   14 cars   460 tons tare   490 tons full

to Newton Abbot   12 cars   393 tons tare   420 tons full

Dist. miles		Actual m. s.	Speeds m.p.h.
0·0	PADDINGTON	0 00	—
9·1	Southall	12 20	61
18·5	SLOUGH	21 00	71
24·2	Maidenhead	25 51	71
31·0	Twyford	31 31	70
36·0	READING	35 58	—
		p.w.s.	
44·8	Aldermaston	46 37	66
53·1	NEWBURY	54 10	64/66
61·5	Hungerford..	61 49	62/67
70·1	Savernake	70 04	50
81·1	Patney	79 14	82
		p.w.s.	
91·4	Edington	88 20	
94·6	*Heywood Rd. Junc.*	91 08	65/72
102·3	*Blatchbridge Junc.*	97 48	62/71
108·3	*Brewham Box*	103 11	60
115·1	Castle Cary	108 49	77/55
125·5	Somerton	117 00	82/75
138·0	*Cogload Box*	126 25	84 (max)
142·7	TAUNTON..	130 18	70
149·8	Wellington	136 45	60
153·6	*Whiteball Box*	141 37	37
158·6	Tiverton Junc.	146 18	77
166·3	Silverton	153 01	80 (max)
173·5	EXETER	159 14	
		p.w.s.	
182·0	Starcross	170 50	
185·7	Dawlish	175 08	
188·5	Teignmouth	178 44	
193·7	NEWTON ABBOT	186 15	

Schedule: to Exeter 167½ min.

to Newton 192 min.

83·1 m.p.h. A maximum speed of 89 m.p.h. was attained at Dauntsey, and the net average speed from start to stop was 77·3 m.p.h.

Although the King on many occasions proved themselves to be very fast runners, it was as heavy-load engines that they were principally in demand, on the West of England and on the Birmingham services. One of the hardest post-war runs was made during a series of dynamometer car tests run in 1955 prior to the acceleration of the Cornish Riviera Express to its pre-war schedule of four hours, non-stop from Paddington to Plymouth. This required a passing time of 167½ min. over the 173·5 miles from Paddington to Exeter, and with a maximum load train of 14 coaches, to Heywood Road Junction (for Westbury) and 12 coaches beyond. Engine No. 6013 *King Henry VIII* was worked hard throughout to see what margin, if any, existed in these conditions.

The result was the very fine run tabulated. To Heywood Road there was not a great improvement on the best standards of pre-war running. But afterwards when some easing of the effort was traditional, the engine continued to be steamed very hard, and in consequence some very fast times were achieved to both Taunton and Exeter. The time to the latter place, 159 min. 14 sec., was 8¼ min. clear inside schedule, and the net time was 155¾ min. Even with such splendid standards of performance it was not possible to attempt unaided working of 12-coach trains over the tremendous gradients of the South Devon line. The maximum tonnage permitted to an unassisted King was 360 tons tare, and so with the full 12-coach train a stop had to be made at Newton Abbot for a bank engine. The start to stop run of 186¼ min. from Paddington with its net time of 181 min. represented a magnificent piece of locomotive work, and fitly rounds off this account of Kings and Castle performance.

"The Merchant Venturer"

Hauled by double-chimneyed King, The "Merchant Venturer" emerges from the short St Annes Park tunnel in the last mile into Bristol

[*G. F. Heiron*

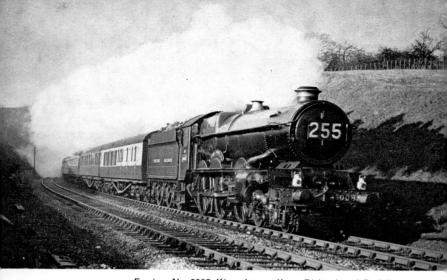

Engine No 6008 *King James II* on Birkenhead-Paddington
express leaving White House Farm tunnel near Beaconsfield, February 1949   *[J. F. Russell-Smith*

State-owned Kings

The 9 30am Paddington-Plymouth express near Rattery
summit, South Devon, in April 1956 with engine No 6025
King Henry V   *[D. J. Fish*

The seven-coach down "Bristolian" hauled by engine
No 7024 *Powis Castle*, approaching Chippenham at 85mph
[*Kenneth H. Leech*

Celebrated Named Trains

The down "Pembroke Coast Express" in Sonning Cutting,
hauled by No 7009 *Athelney Castle*; July 1955
[*M. W. Earley*

Kings *en echelon*, waiting at Newbury Racecourse station with up race specials for Paddington: No 6005 *King George II* and No 6000 *King George V*   [*M. Pope*

Double-chimneyed Kings

Engine No 6010 *King Charles I* on Saturdays-only up Cornish express (12-coach load) approaching Whiteball summit.
                                                    [*Kenneth H. Leach*

CHAPTER TWELVE

Personalia of the Castles and Kings

UNTIL the late 1930's the Great Western Railway made few exceptions to its scheme of systematic naming for the express passenger and mixed traffic engines. Locomotives of the Castle class were named after castles—some large and famous one like Windsor and Warwick, some great ruins like Caerphilly and Chepstow, and others perhaps not so well known, but nevertheless attractive in the euphony of the name itself, such an *Llandovery Castle*, *Llanstephan Castle*, *St. Donats Castle*. For the twenty engines numbered from 5043 upwards a further series of beautiful names was chosen, though by this time many of the castles could hardly be called well known. At about the same time too, the new lightweight 4-4-0s of the 3200 class were being given names of Earls associated with the G.W.R. It was, however, soon felt that the latter names should be more appropriately carried on express passenger engines than on a class designed particularly for the Cambrian line, and so, as the engines concerned passed through Swindon for overall, these names were transferred to the 5043-5062 series of Castles, and the original names were held in reserve for future engines of the Castle class. A further engine, No. 5063, was named *Earl Baldwin*, after the former Prime Minister.

In the general reshuffle of names two other engines were involved, 5064 *Tretower Castle* and 5065 *Upton Castle*; these at once took two of the displaced names and became *Bishop's Castle* (ex-5053) and *Newport Castle* (ex-5058). Three, however, *Sudeley Castle*, *Tenby Castle* and *Thornbury Castle*, were not used until 1949. The remainder of the displaced ones reappeared on the 5068-5082 and 5093-5097 series, built in 1938-9. But when the war came, and with it the Battle of Britain, the intense feelings of patriotism and pride in the achievements of the Royal Air Force found its expression on the G.W.R. in the re-naming of ten Castle class engines after types of aircraft, the names of which were then on everyone's lips. And so once again the castles Clifford, Compton, Cranbrook, Denbigh, Devizes, Drysllwyn, Eastnor, Lamphey, Lydford, Ogmore, Penrice and Powis—or rather the nameplates—went back into the stores at Swindon, to make way for *Spitfire*, *Hurricane*, *Blenheim*, *Hampden*, *Wellington*, *Gladiator*, *Fairey Battle*, *Beaufort*, *Lysander*, *Defiant*, *Lockheed Hudson* and *Swordfish*.

81

E

In 1946, when the Hawksworth Castles were constructed, eight of the twice-displaced names were brought out for the third time, but since then—and this is surely a record—two of them, *Denbigh Castle* and *Ogmore Castle*, have been displaced for the third time. Engine No. 7001 was renamed *Sir James Milne*, in honour of the last General Manager of the G.W.R., and No. 5066 *Sir Felix Pole* while after nationalisation No. 7007 was re-named to commemorate the railway itself, *Great Western*, and carried the coat of arms on the splashers beneath the nameplates. The name *Great Western* is a revival of an old one carried by one of the broad gauge 8 ft. singles, and later by a standard gauge Dean 7 ft. 8 in. single. After the scrapping of the latter engine in May, 1909, the name was not used again until it was bestowed upon No. 7007 in 1948.

Two other names outside the realm of Castles are those borne by 5069 and 5070, named after the two great pioneer engineers of the G.W.R. Prior to this there had been several engines named after Brunel and Gooch, but on none previously had the names been more splendidly rendered. In some cases the name was just *Gooch* and, in one case *Sir Daniel*; but on these new Castles the names were verily emblazoned in full, *Isambard Kingdom Brunel*, and *Sir Daniel Gooch*.

In the 1946 series of Castles, although four of the names displaced in 1940 were not used again—Cranbrook, Drysllwyn, Penrice and Powis—there was one that was quite new, *Elmley Castle*, on No. 7003, and of this series too, No. 7000 was named *Viscount Portal* after the Chairman of the Company. Engine No. 7005 *Lamphey Castle* was subsequently renamed *Sir Edward Elgar*. The 1948 batch, 7008-7017, all carried entirely new names, and then it seemed unlikely that the thrice-displaced *Denbigh Castle* and *Ogmore Castle*, together with the four that were twice displaced, would be used again. As it turned out they were all used on the 1949 and 1950 batches. Of the 1948 series, Nos. 7010-7013 were turned out new in the experimental light green livery, though this was not perpetuated. Five others, 4089 *Donnington Castle*, 4091 *Dudley Castle*, 5010 *Restormel Castle*, 5021 *Whittington Castle* and 5023 *Brecon Castle* were also re-painted in light green.

Among these engines in the light green livery *Brecon Castle* carried for a time a new crest in which the British Lion was incorporated with various insignia symbolical of railways. This crest recalled the small variations in style of painting that have taken place during the lifetime of the Castles and the Kings. With the construction of the first Castles in 1923, the pre-war style of painting was revived—the famous Brunswick Green, with black and orange lining—and the tenders carried the words Great Western on each side of a crest that included the G.W.R. coat of arms surrounded by a garter. At a later stage this device was replaced by the coat of arms alone, and then from 1934 onwards, the appearance of Great Western tenders became much plainer, as a small monogram including the letters G.W.R. in block characters within a circle replaced the full name and the coat of arms. Since the war the coat of arms was revived, but flanked by the initials G.W. instead of the full name. Yet another stage was reached when the long-familiar name was replaced by the inscription 'British Railways,' though still in the Great Western style of painting.

Quite apart from the experimental liveries there have been six styles of tender painting on the Castles and Kings, for one must include also the plain unlined green of wartime. Four of the Kings were, in 1948, painted in the experimental dark blue, with red, cream and grey lining, and having the name British Railways in sans serif characters on the tender. To one accustomed to the Brunswick green they looked a little strange, perhaps even garish to some eyes; for my own part I was glad when this was abandoned. But while the design of the engines themselves remained unchanged for more than 40 years, their tenders bore no less than eight different styles of painting !

The Castles, as the standard passenger engines of the line, were stationed at many widely separated depots. As one might expect, Old Oak Common had more than any other shed—usually over 30, and there were studs of a dozen or so at Bristol (Bath Road), Plymouth (Laira), Cardiff, Newton Abbot, Swansea (Landore), and Wolverhampton (Stafford Road). There were smaller contingents at Exeter, Swindon, Worcester and Shrewsbury, while taking one particular date as typical, January, 1947, there was one at Reading (4085), two at Taunton (5003 and 5077), three at Gloucester, and one at Hereford. The latter was the celebrated 4079 *Pendennis Castle*. With one or two exceptions the Great Western never allocated engines in approximate harmony with their names; thus one found *Rougemont Castle* not at Exeter but at Cardiff, *Shrewsbury Castle* at Plymouth, and *Llantilio Castle* at Newton Abbot. The same applied to the first two engines which were renamed after famous regiments: No. 4016 *Knight of the Golden Fleece*, which became *Somerset Light Infantry*, and 4037 *Queen Philippa* which was renamed *The South Wales Borderers*. Both these engines were rebuilt from Stars. Now Cardiff shed had a reputation for making its express locomotives outshine all the others in the spotlessness of their turnout; if they could do this with *Rougemont Castle, Trematon*

Castle and *Earl Cawdor*, what might they not have done with *The South Wales Borderers*? But unfortunately she was not stationed at a Welsh shed. The outstanding exception was No. 5017 which was named *The Gloucestershire Regiment 28th/61st*, in honour of the glorious part played by that regiment in the Korean War. Engine No. 5017 was stationed at Gloucester.

The Castles surely held a record immunity for mishap. The only one to be involved in a serious accident was No. 4091 *Dudley Castle*, which, in the early hours of July 2nd, 1941, crashed head-on into the engine of a down freight train on the diamond crossing of Dolphin Junction, near Slough. The up Plymouth express worked by No. 4091 was to be diverted to the relief line. The signalman thought the down freight had come to a stand, and lowered the signals for the express to cross over; far from being safely at a stand, the freight was over-running its signals and a severe collision resulted in five passengers being killed. On two other occasions Castles had narrow escapes, as when the midnight express from Paddington to Birkenhead, drawn by No. 4088 *Dartmouth Castle*, was fouled and partially wrecked by a goods train derailment at Appleford Crossing near Didcot, on November 13th, 1942. No. 4088 and the two leading vans were, however, clear before the over-turning wagons struck the train. No. 5061 *Earl of Birkenhead* was concerned in a most alarming night collision at a level crossing on the Chester line; the train was travelling at between 70 and 75 m.p.h. when a car was struck, and Driver Jack Edwards, of Shrewsbury, told me afterwards how the engine rocked wildly from side to side, but with the greatest good fortune kept the rails.

The Kings were not so lucky in this respect. No. 6007 *King William III* was involved in the first accident on the G.W.R. that could be termed serious since the derailment at Loughor in 1904 when three lives were lost. The smash at Shrivenham on January 15th, 1936, was primarily due to the breakage of a drawhook on an up mineral train; five wagons and the brake van became detached and were left in the Marston East-Shrivenham section. The signalman at Shrivenham, being somewhat pre-occupied, did not notice the freight train was incomplete, gave "out of section" to the box in rear, and then accepted the up Penzance sleeper. The express, hauled by No. 6007 and running under clear signals, hit the detached portion of the freight train at practically full speed; but although the collision was violent, there were only two deaths—the driver of 6007 and one passenger. A far more serious accident was that to the corresponding down express on November 4th, 1940, when the driver, on leaving Taunton, assumed he was running on the fast line, whereas he was actually on the relief. Near Norton Fitzwarren he was overtaken by the 12.50 a.m. newspaper express from Paddington running on the adjacent fast line and realised his error; but it was too late, and at the point where the four tracks from Taunton converge into two, this heavy sleeping car express packed with passengers and travelling at 40 m.p.h., ran through the catch points with disastrous results. The engine, No. 6028 *King George VI*, overturned on to fairly soft ground and was not seriously damaged, but casualties in the first six coaches were heavy. In all 27 persons were killed and 56 seriously injured. Regrettable though this derailment was, its occurrence served to emphasise the remarkable immunity from accident that the Great Western Railway had earned, partly due to the universal application on its main lines of automatic train control apparatus. This was indeed the first accident on the G.W.R. for over 40 years in which more than five persons had been killed.

In bringing this account towards its close, I must mention the experimental fitting of streamlined fairings to engines 6014 *King Henry VII* and 5005 *Manorbier Castle* early in 1935. In addition to the air-smoothing of forward projections an attempt was made to fill in the spaces immediately behind the boiler mountings, which spaces are prone to set up eddies, and so increase the air resistance at speed. The resulting effect, though queer, would not have looked too bad if it had not been for the hemi-spherical appurtenance fitted on to the smokebox front; as it was, the effect was hideous. These streamlined fairings were removed some time ago.

A naming ceremony was arranged when No. 4009 *Shooting Star* was renamed *Lloyds*, after the world-famous firm of shipping insurance agents, and similar ceremonies were performed for each of the regimental engines 4016 and 4037. But the ceremony that appealed so strongly to all locomotive men was that performed at Paddington on Friday, October 29th, 1948, by the President of the Institution of Mechanical Engineers, when the latest Castle then built, No. 7017, was renamed *G. J. Churchward*. Thus the 43 years were spanned since Churchward, in constructing the first Great Western four-cylinder simple express engine, laid the foundations of modern British locomotive practice; he developed this design till, in the Princes and Princesses of 1913-1914, it may be termed his masterpiece. On such a foundation Collett, building the Castles and Kings in the established Swindon tradition, proved himself no less a master hand; and then in later times F. W. Hawksworth, by his own additions to the stud, has continued the great tradition. On November 15th, 1950 Her Majesty the Queen visited Swindon works and on this occasion she named the latest Castle to be built, No. 7037 *Swindon*.

Appendices

Appendices

NAMES OF CASTLE CLASS LOCOMOTIVES NOT LISTED IN EARLIER CHAPTERS

5013-5042 SERIES

5013	Abergavenny Castle	5028	Llantilio Castle
5014	Goodrich Castle	5029	Nunney Castle
5015	Kingswear Castle	5030	Shirburn Castle
5016	Montgomery Castle	5031	Totnes Castle
5017	St. Donats Castle*	5032	Usk Castle
5018	St. Mawes Castle	5033	Broughton Castle
5019	Treago Castle	5034	Corfe Castle
5020	Trematon Castle	5035	Coity Castle
5021	Whittington Castle	5036	Lyonshall Castle
5022	Wigmore Castle	5037	Monmouth Castle
5023	Brecon Castle	5038	Morlais Castle
5024	Carew Castle	5039	Rhuddlan Castle
5025	Chirk Castle	5040	Stokesay Castle
5026	Criccieth Castle	5041	Tiverton Castle
5027	Farleigh Castle	5042	Winchester Castle

5043-5067 SERIES

	Original Name		Later Name
5043	Barbury Castle	..	Earl of Mount Edgcumbe
5044	Beverston Castle	..	Earl of Dunraven
5045	Bridgwater Castle	..	Earl of Dudley
5046	Clifford Castle	..	Earl Cawdor
5047	Compton Castle	..	Earl of Dartmouth
5048	Cranbrook Castle	..	Earl of Devon
5049	Denbigh Castle	..	Earl of Plymouth
5050	Devizes Castle	..	Earl of St. Germans
5051	Drysllwyn Castle	..	Earl Bathurst
5052	Eastnor Castle	..	Earl of Radnor
5053	Bishops Castle	..	Earl Cairns
5054	Lamphey Castle	..	Earl of Ducie
5055	Lydford Castle	..	Earl of Eldon
5056	Ogmore Castle	..	Earl of Powis
5057	Penrice Castle	..	Earl Waldegrave
5058	Newport Castle	..	Earl of Clancarty
5059	Powis Castle	..	Earl St. Aldwyn
5060	Sarum Castle	..	Earl of Berkeley
5061	Sudeley Castle	..	Earl of Birkenhead
5062	Tenby Castle	..	Earl of Shaftesbury
5063	Thornbury Castle	..	Earl Baldwin
5064	Tretower Castle	..	Bishop's Castle
5065	Upton Castle	..	Newport Castle
5066	Wardour Castle	..	Sir Felix Pole
5067	St. Fagan's Castle	..	St. Fagan's Castle

* renamed *The Gloucestershire Regiment 28th/61st*

5068-5097 SERIES

5068	Beverston Castle				
5069	Isambard Kingdom Brunel				
5070	Sir Daniel Gooch				
5071	Clifford Castle	..	..	*Renamed*	Spitfire
5072	Compton Castle	..	..	,,	Hurricane
5073	Cranbrook Castle		..	,,	Blenheim
5074	Denbigh Castle	..	..	,,	Hampden
5075	Devizes Castle	..	..	,,	Wellington
5076	Drysllwyn Castle		..	,,	Gladiator
5077	Eastnor Castle	..	..	,,	Fairey Battle
5078	Lamphey Castle	..	..	,,	Beaufort
5079	Lydford Castle	..	..	*Renamed*	Lysander
5080	Ogmore Castle	..	..	,,	Defiant
5081	Penrice Castle	..	..	,,	Lockheed Hudson
5082	Powis Castle	..	..	,,	Swordfish
5083	Bath Abbey	..	..	*Rebuilt from*	4063
5084	Reading Abbey	..	..	,, ,,	4064
5085	Evesham Abbey	..	..	,, ,,	4065
5086	Viscount Horne	..	..	,, ,,	4066
5087	Tintern Abbey	..	..	,, ,,	4067
5088	Llanthony Abbey		..	,, ,,	4068
5089	Westminster Abbey		..	,, ,,	4069
5090	Neath Abbey	..	..	,, ,,	4070
5091	Cleeve Abbey	..	..	,, ,,	4071
5092	Tresco Abbey	..	..	,, ,,	4072
5093	Upton Castle				
5094	Tretower Castle				
5095	Barbury Castle				
5096	Bridgwater Castle				
5097	Sarum Castle				

5098 CLASS

5098 Clifford Castle
5099 Compton Castle
7000 Viscount Portal
7001 Denbigh Castle *Renamed* Sir James Milne
7002 Devizes Castle
7003 Elmley Castle
7004 Eastnor Castle
7005 Lamphey Castle *Renamed* Sir Edward Elgar
7006 Lydford Castle

7007   Ogmore Castle *Renamed* **Great Western**
7008   Swansea Castle
7009   Athelney Castle
7010   Avondale Castle
7011   Banbury Castle
7012   Barry Castle
7013   Bristol Castle
7014   Caerhays Castle
7015   Carn Brea Castle
7016   Chester Castle
7017   G. J. Churchward
7018   Drysllwyn Castle
7019   Fowey Castle
7020   Gloucester Castle
7021   Haverfordwest Castle
7022   Hereford Castle
7023   Penrice Castle
7024   Powis Castle
7025   Sudeley Castle
7026   Tenby Castle
7027   Thornbury Castle
7028   Cadbury Castle
7029   Clun Castle
7030   Cranbrook Castle
7031   Cromwell's Castle
7032   Denbigh Castle
7033   Hartlebury Castle
7034   Ince Castle
7035   Ogmore Castle
7036   Taunton Castle
7037   Swindon

The class was completed by six conversions :-

4000	North Star..	*converted from Star class*
100A1	Lloyds	*formerly* 4009 Shooting Star
4016	The Somerset Light Infantry (Prince Albert's)	
		formerly Knight of the Golden Fleece
4032	Queen Alexandra	*converted from Star class*
4037	The South Wales Borderers	*formerly* Queen Philippa (Star *class*)
111	Viscount Churchill ..	*rebuilt from* The Great Bear

DIAGRAMS OF
LOCOMOTIVES AND LOCOMOTIVE FOOTPLATES

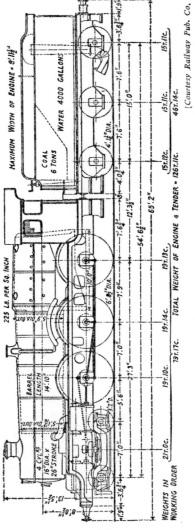

[Courtesy Railway Pub. Co.

Castle Class

MAXIMUM WIDTH OF ENGINE = 8'.11½"

WATER 4000 GALLONS

COAL 6 TONS

225 LB. PER SQ. INCH

BARREL LENGTH 14'.10"

4 CYL.^s 16"DIA.x 26"STROKE

TOTAL WEIGHT OF ENGINE & TENDER = 126T.11C.

WEIGHTS IN WORKING ORDER

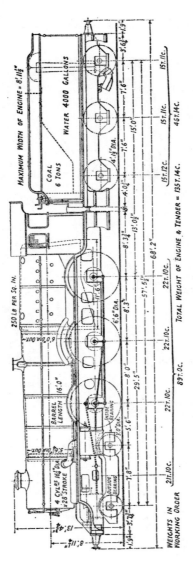

King Class

CASTLE CLASS FOOTPLATE

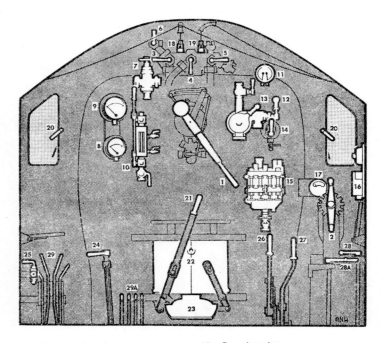

1. Regulator handle
2. Reversing gear handle
3. Exhaust injector live steam valve
4. Exhaust steam injector auxiliary steam valve
5. Right-hand injector live steam valve
6. Steam heating valve
7. Steam heating pressure regulator
8. Steam heating pressure gauge
9. Boiler steam pressure gauge
10. Water level gauge
11. Duplex vacuum brake gauge
12. Ejector steam valve
13. Brake application lever
14. Blower valve
15. Hydrostatic lubricator
16. Audible signalling and automatic train control apparatus

17. Speedometer
18. Signal whistle valve
19. Brake whistle valve
20. Screen wipers
21. Firedoor operating lever
22. Double firedoors
23. Firehole half door
24. Exhaust injector water regulator
25. Coal watering cock
26. Cylinder cocks control lever
27. Front sanding lever
28. Rear sanding lever
28a. Rear sanding lever alternative position
29. Air damper levers
29a. Air damper levers alternative position

KING CLASS FOOTPLATE

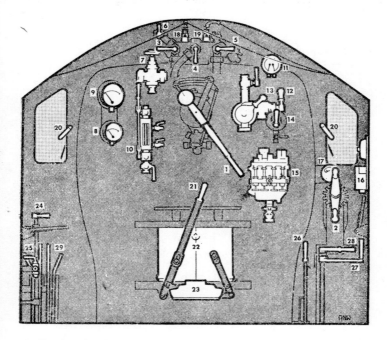

1. Regulator handle.
2. Reversing gear handle.
3. Exhaust injector live steam valve.
4. Exhaust steam injector auxiliary steam valve
5. Right-hand injector live steam valve.
6. Steam heating valve.
7. Steam heating pressure regulator.
8. Steam heating pressure gauge.
9. Boiler steam pressure gauge.
10. Water level gauge.
11. Duplex vacuum brake gauge.
12. Ejector steam valve.
13. Brake application lever.
14. Blower valve.
15. Hydrostatic lubricator.
16. Audible signalling and automatic train control apparatus.
17. Speedometer.
18. Signal whistle valve.
19. Brake whistle valve.
20. Screen wipers.
21. Firedoor operating lever.
22. Double firedoors.
23. Firehole half door.
24. Exhaust injector water regulator.
25. Coal watering cock.
26. Cylinder cocks control lever.
27. Front sanding lever.
28. Rear sanding lever.
28a. Rear sanding lever alternative position.
29. Air damper levers.
29a. Air damper levers alternative position.

CASTLES AND KINGS—LEADING DIMENSIONS

Type	Original Castles	Hawksworth Castles	Kings
Cylinders (4) diameter ..	16″	16″	16¼″
stroke ..	26″	26″	28″
Coupled Wheels, diameter	6′ 8½″	6′ 8½″	6′ 6″
Bogie Wheels, diameter..	3′ 2″	3′ 2″	3′ 2″
Wheelbase coupled ..	14′ 9″	14′ 9″	16′ 3″
Wheelbase total engine ..	27′ 3″	27′ 3″	29′ 5″
Heating surface—			
Firebox	162·7 sq. ft.	163·5 sq. ft.	193·5 sq. ft.
Tubes and Flues ..	1,857·7 sq. ft.	1,799·5 sq. ft.	2,007·5 sq. ft.
Total Evaporative ..	2,020·4 sq. ft.	1,963 sq. ft.	2,201·0 sq. ft.
Superheating surface ..	262·6 sq. ft.	295 sq. ft.	313 sq. ft.
Total combined heating			
surface	2,283 sq. ft.	2,258 sq. ft.	2,514 sq. ft.
Firegrate area	29·4 sq. ft.	29·4 sq. ft.	34·3 sq. ft.
Working pressure,			
per sq. in.	225 lb.	225 lb.	250 lb.
Adhesion Weight.. ..	58·8 tons	58·8 tons	67·5 tons
Weight of engine in			
working order	79·8 tons	79·8 tons	89 tons
Water capacity of tender	4,000 gal.	4,000 gal.	4,000 gal.
Coal capacity of tender ..	6 tons	6 tons	6 tons
Weight of engine and			
tender	126·5 tons*	127·3 tons†	135·7 tons
Tractive effort			
(at 85 per cent. b.p.) ..	31,625 lb.	31,625 lb.	40,300 lb.

* With "Collett"-type tender.
† With "Hawksworth"-type straight-sided tender.